A Review Of Marking And Individual Recognition Techniques For Amphibians And Reptiles

John W. Ferner

Department of Biology
Thomas More College
Crestview Hills, Ky 41017

and

Adjunct Curator of Herpetology
Cincinnati Museum Center

2007

SOCIETY FOR THE STUDY OF AMPHIBIANS AND REPTILES

HERPETOLOGICAL CIRCULAR NO. 35

Published February 2007
© 2007 Society for the Study of Amphibians and Reptiles

John J. Moriarty, *Editor*
3261 Victoria Street
Shoreview, MN 55126 USA
frogs@umn.edu

Single copies of this circular are available from the Publications Secretary, Breck Bartholomew, P.O. Box 58517, Salt Lake City, Utah 84158-0517, USA. *Telephone and fax:* (801) 562-2660 *E-mail:* ssar@herplit.com. A list of other Society publications, including *Facsimile Reprints in Herpetology, Herpetological Conservation, Contributions to Herpetology*, and the *Catalogue of American Amphibians and Reptiles*, will be sent on request or can be found at the end of this circular.

Membership in the Society for the Study of Amphibians and Reptiles includes subscription to the Society's technical *Journal of Herpetology* and semi-technical *Herpetological Review*. Currently, Regular dues are $60.00 ($30.00 for students), Plenary $80.00 (includes JH, HR, and CAAR), and Institutional subscriptions are $115.00. Non-USA members and institutions can add $35.00 for optional air mail delivery. Subscription to the *Catalogue of American Amphibians and Reptiles* is an additional $20.00. All inquiries about SSAR membership or subscriptions should be addressed to the SSAR Membership Office, P.O. Box 58517, Salt Lake City, Utah 84158-0517, USA. *Telephone and fax:* (801) 562-2660 *E-mail:* ssar@herplit.com.

Front Cover: Lizard: European Wall Lizard (*Podarcis muralis*) , photo by John Ferner and Jodi Carter, paint symbol and toe clipping after Tinkle (1967).
Snake: Queen Snake(*Regina septemvittata*), , photo by Kent Bekker, ventral scute clip after Brown and Parker (1976b).

ISBN 0-916984-68-0

Table of Contents

Preface	ii
Larval Amphibians	1
Frogs and Toads	3
Salamanders	14
Caecilians	20
Turtles	20
Lizards	29
Snakes	37
Crocodilians	46
Passive Integrated Transponder (PIT) Tags	47
Radio Transmitter Tagging	51
Literature Cited	60

Preface

Since my earlier review of marking techniques (Ferner, 1979) there have continued to be numerous modifications and analysis of existing methods as well as the development of new procedures. Some of the older techniques are used much less often. The use of radioactive tags and various mutilation procedures are some examples. On the other hand, major advancements in biotelemetry and the new technique of using Passive Integrated Transponder (PIT) tags have added much needed flexibility in the choice of marking methods. More field and behavioral studies which require marking are now being conducted than in the past and often investigators are in need of marking techniques and noninvasive ways of identifying individuals with very specific qualifications. In addition, ethical issues relative to certain techniques are now more commonly considered such as with amphibian toe clipping which is discussed in a separate section. It is for these reasons that I am writing this updated review of marking and individual recognition techniques for amphibians and reptiles.

Partial reviews of marking and identification techniques for amphibians and reptiles have been published in recent years. These include reviews of techniques for all animals in the book edited by Stonehouse (1978), for amphibians by Donnelly et al. (1994), for reptiles by Plummer and Ferner (in press) and for turtles by Plummer (1979).

Some criteria for an ideal mark or tag are as follows:

1. It should not affect the survivorship or behavior of the organism (Ricker, 1956).
2. It should allow the animal to be as free from stress or pain as possible (Lewke and Stroud, 1974).
3. It should identify the animal as a particular individual, if desirable.
4. It should last indefinitely (Ricker, 1956).
5. It should be easily read and/or observable (Lewke and Stroud, 1974).
6. It should be adaptable to organisms of different sizes (Lewke and Stroud, 1974).
7. It should be easy to use in both laboratory and field, and use easily obtained material at minimal cost (Lewke and Stroud, 1974).

There are, of course, no techniques that satisfy all of these desired criteria to the fullest. Selection of a technique, then, will require deciding which of the criteria are most important for any particular study. Often, two or more marks are used in order to meet more of the criteria (e.g., toe clipping and dorsal paint symbols). Also, a technique described for use with one group of herpetofauna might be adapted for use with a new group and therefore one might benefit by scanning other portions of this guide. There is a need for more standardization of marking techniques in comparative studies or those that might be carried on by other investigators in future years.

I have given most space to the more commonly used techniques and to the newer ones which may not have yet stood the test of time. When available, information on the advantages and disadvantages of each technique has been given.

For the most part, this guide attempts to give information adequate for use of techniques without consulting the original source. However, once a technique has been tentatively selected it seems advisable to consult the original source if your time and facilities permit. The use of very complex techniques, such as those using radiotelemetry or PIT tags which are covered in separate chapters, require careful review of the original sources. These more involved techniques usually need to be adapted for each particular study. Discussion of marking techniques for the more exotic reptiles, such as sea turtles and crocodilians, is limited probably due to the fact that the existing marking techniques are working well and have been used for an extended time period with little modification. Satellite tracking of these larger reptiles uses some of the most advanced technologies. Sources for materials used with various techniques are given as mentioned in the original publication and no longer may be available. If I found more updated information on a supplier, that is indicated in the text. An additional shortcoming of this review is that many valuable techniques have no doubt been omitted, primarily due to my not having uncovered them in the literature or being able to locate a copy.

The assistance I received in finding references with the first edition of this review in 1979 is acknowledged once again. These included R. E. Ashton, R. W. Barbour, R. K. Farrell, H. S. Fitch, C. J. McCoy, S. J. Gorzula, M. V. Plummer and R. G. Zweifel. Many of the drawings which Karen Brauer redrew from original figures were used again in this edition. The support of my colleagues at Thomas More College, Bill Bryant and Chris Lorentz, and at the Cincinnati Museum Center, Jeff Davis, Paul Krusling, and Rob Stuebing has also been helpful. Thanks are also extended to my wife, Jodi, for her support and suggestions on the preparation of this review.

John W. Ferner
Crestview Hills, KY

LARVAL AMPHIBIANS

Due to their small size and delicate tissues, amphibian larvae require special marking techniques. A traditional method, used by Turner (1960), simply involves clipping notches from the tail fin of tadpoles. Guttman and Creasy (1973), however, reported that the fin clipping technique causes a higher mortality than staining tadpoles with Neutral Red. Herreid and Kinney (1966) first used this staining technique by placing *Rana sylvatica* tadpoles in shallow pans with a 0.05% solution of Neutral Red and pond water for 30 minutes. In laboratory studies they found that dyed and undyed tadpoles survived equally well. Herreid and Kinney (1966) found this staining technique adequate for a mark-recapture study, although the dye usually faded within one week. Neutral Red has also been used successfully to stain tadpoles of *Rana clamitans* which were immersed for 3 hours in concentrations of 1 part stain to 25,000 or 50,000 parts of pond water (Guttman and Creasy, 1973). These tadpoles developed a bright crimson venter, had an immediate mortality rate of 8.7% (N-567) and retained the stain for at least 10 days. Travis (1981) reports that *Hyla gratiosa* tadpoles stained with neutral red grew more slowly than the control group at low density, but at high density there was no significant difference in growth rate. All staining methods appear to be time limited.

Batch marking of *Bufo boreas* metamorphs with oxytetracycline as reported by Muths et al. (2000) may be an appropriate technique for some studies such as those translocating individuals to new habitats in restoration projects. This technique does require the sampling of tissue (toe clip) in the fully metamorphosed adult in order to detect any microscopic fluorescence from the tetracycline label. Taylor and Deegan (1982) inject a fluorescent pigment into *Rana clamitans* (Gosner stage 25) dermis using a compressed-air spray gun resulting in minimal mortality of 3% over a month after marking. Short-term batch marking of *Ambystoma opacum* and *A. maculatum* larvae with 24-sodium (half-life = six days) by Shoop (1971) was successful, but technology needs and required permits may make this technique impractical for most studies. Cecil and Just (1978) injected three recognizably different colors of acrylic polymers into the ventral or dorsal tail fins of *Rana catesbieana* tadpoles using a 2.5 cc syringe with an 18 to 26 gauge needle. A small dot of color (1 to 3 mm diameter) was created on each fin with a resulting six labels that could be used with different collections of tadpoles for laboratory experiments. The tags were retained for five to six months in the 350 tadpoles tested in the lab with no known adverse effects and for up to two years for 8000 tadpoles in the field. The acrylic dye tag was resorbed at metamorphosis with no known impact on the frogs (Cecil and Just, 1978).

Regester and Woosley (2005) used visible implant fluorescent elastomer (VIE) (Northwest Marine Technologies, Inc., Shaw Island, Washington, USA) to batch mark egg masses of *Ambystoma maculatum*. Yellow VIE was injected into the jelly matrix of the egg mass with a 33 mm, rounded top, 18-gauge needle. Two test groups were injected in both the laboratory and field, one with 25 mm of VIE into the inner jelly layer (IM-VIE) and the second with 10 mm into the outer jelly layer (OM-VIE). Controls had the needle inserted into either the inner or outer jelly layers with no VIE injected. The laboratory sample retained identifiable marks through hatching (>35days) and in the field OM-VIE marks remained for 83% of the development days and for the IM-VIE for 97% of the development days. Regester and Woosley (2005) reported no effect of VIE use on the survival or hatching of the salamander embryos and the developmental stage and body size of the larvae were normal.

A permanent marking method has been described for *Rana* and *Ambystoma* larvae by Seale and Boraas (1974). This technique uses a subcutaneous injection of an organic dye-organic solvent mixture into the tail fin. The authors, by trial and error, determined the ideal ratio of mineral oil to petroleum jelly of 21:20 (by weight). If an excess of either component is used, undulations of the tail fin may cause the ejection of the marks. The ratio may vary depending on the viscosity of the commercial products used, but once determined, the same ratio can be used for all species. Organic biological stains ("Oil Red O" and "Oil Blue M" from Matheson, Coleman and Bell) were used to dye the oil. Variations in color and placement of marks were used to identify groups of individuals marked on specific dates. Marks were placed in the dorsal or ventral tail fin cavity adjacent and parallel to tail musculature by using a 22-guage hypodermic syringe. The needle was inserted about 0.6 cm into the cavity and then removed while pressure was applied with the syringe plunger; this procedure left a 0.5 cm line of dye. No mortality, infection, impairment of mobility or retardation of growth were found as a result of this marking technique. Any individuals injured during the marking process were removed before release and caution was particularly needed to prevent injury to the gills of larval *Ambystoma*. During metamorphosis, this organic dye-organic solvent mark is resorbed with the tail with no discernible ill effects.

An additional method of marking salamander larvae, using fine grained (20 *microns*) fluorescent pigments, was used by Ireland (1973). Four colors of these pigments (solid solutions of fluorescent dyes in a melamine-sulfonamide-formaldehyde resin) were mixed with acetone to make a paste. This paste was then administered to the larvae on their mid-dorsal surface with a blunt probe which had been heated with a propane torch. The probe burned through the outer epithelial layers, leaving a small scar (1 mm). The epithelium regenerated within 15 days and incorporated the fluorescent pigments. These tags were then observed by placing the larvae in a petri dish containing water, setting the dish in a black box and then exposing them to long wave ultraviolet light. In the laboratory, 50% of the *Eurycea multiplicata* tested had lost their fluorescent tag in 70 days and 47% of the *Ambystoma annulatum* had done so. It appeared as though loss of the tags in the field was less than this. No detrimental effects of this tag on the larvae were observed.

After anaesthetizing tadpoles of *Rana temporaria*, *R. esculenta* and *R. lessonae* in a 0.02% solution of TMS (tricaine methanesulfonate), Anholt et al. (1998) rinsed them and marked them by injecting a biologically inert fluorescent elastomer under the skin and above the musculature in the tail. Using a 1 ml tuberculin syringe and a No. 20 needle the colored elastomer (Northwest Marine Technology, Inc.) mark about 0.2 mm wide and 1 mm long was injected while observing under a stereomicroscope. Using the five colors available up to 80 individuals could be marked differently using no more than three injections each. The marks were not found to impact the tadpole behavior or survival, but mark retention was around 85%.

Rice et al. (1998) describe a technique for tying plastic field flagging tags onto anesthetized (0.050 g tricaine methanesulfonate per L dechlorinated water) *Rana catesbeiana* using a single stitch of thin polyester thread (0.20 mm diameter). This procedure caused minimal mortality and tag loss was low if the thread was knotted tightly. There was no significant impact on growth of the larvae. The technique seems best suited for laboratory studies which would not have concerns about the snagging or any hydrodynamic drag the tag may cause.

In summary, studies of a particular species may find one of the above techniques more applicable than the others. As these new techniques are used with additional species in the future, more information on their relative merits will be available. Donnelly et al. (1994) point out that marking amphibian larvae is difficult and most techniques are date-specific rather than unique individual marks. Lab testing of all techniques is recommended. However, the technique of Seale and Boraas (1974) appears to be the most permanent and least detrimental to amphibian larvae. On the other hand, the staining method of Guttman and Creasey (1973) seems to be a much simpler technique to master. The detrimental effects, complexity, and time limited nature of the other techniques discussed discourages their use.

FROGS AND TOADS

TOE CLIPPING

Although several toe clipping methods are proposed in the literature, that of Martof (1953) is most widely used. The early modification of this technique by Carpenter (1954) has not been cited in recent literature. Martof cut the toes of *Rana clamitans* with scissors, trying to avoid damage to the webbing between them. Little blood loss or loss of swimming power was observed and no regeneration of the digits was found. Figure 1A shows the system of assigning serial numbers to the digits of the feet. No more than two toes were removed from any one foot. The left hind foot indicated units, and for those larger than five, combinations of two digits (the fifth and one other) were excised. For example, clipping the fifth and third toes designated tens and the front feet denoted hundreds. With this arrangement, one can mark 6399 individuals in series; this, the highest number, can be indicated by removing the two outer digits on each foot. The only concern Martof expressed about this plan was that confusion might arise in those were only a single toe on a single hand foot was removed with capture of a frog with an injured foot. He suggested alleviating this problem with a special designation, such as a zero marking by removal of the second and fourth toes on the other hind foot.

A number code proposed by Hero (1989) reduced the number of toes needing to be clipped over that given by Martof (1953) as shown in Figure 1 A and 1B. Hero can mark 124 anurans using two toe-clips and 736 using three toes. Waichman (1992) proposed a numbering system for toe clipping that letters the four limbs (A through D) and number the toes (Figure 1 C) which makes all the combinations usable for two and three toes, with the stipulation of not removing more than two toes for each limb. A list of the possible combinations (Table 1) can be photocopied and used in the field to keep track of the codes which have been used. Waichman's system can mark up to 959 frogs using up to three toes.

One potential problem with toe clipping is the possibility of regeneration. George (1940) found no regeneration in a three-year study of toe clipped *Rana catesbeiana*. Regeneration may be more commonly found in primitive frogs, but is also found in advanced forms such as the Kenyan reed frog, *Hyperolius viridiflavus ferniquei* (Richards, et al., 1975); newly metamorphosed reed frogs completely regenerate amputated toes, including the structurally complex digital pad. Richards, et al. (1975) also noted that ecologists generally avoid toe clipping tree frogs because of their regenerative capabilities. This depends on the species, as Jameson (1957) found only slight toe regeneration in *Hyla regilla* after one year, and Brown and Alcala (1970) reported slow or nonexistent regeneration of toes in *Rana erythraea* during a four-year field study. Breckenridge and Tester (1961) found that clipped toes of *Bufo hemiophrys* regenerated over a period of several months appearing shorter, duller in color and sometimes having a hooked or pointed shape. They found all toe clips could be recognized for at least a year and many for a longer period.

Briggs and Storm (1970) avoided clipping the thumbs of *Rana cascadae* due to their importance in amplexus and usefulness in sexing individuals. This modification was also made by Turner (1960) with *Rana pretiosa* and Dole and Durant (1974), who additionally did not clip the

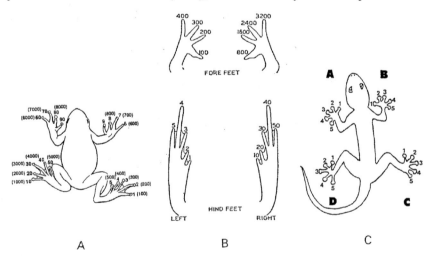

Figure 1. A. Toe clipping code for anurans: when two clips are made on one side of the body, the bracketed number is adopted by the lowest digit (e.g. marking the second and fourth digits on the right foot would produce the number 204 not 402). From Hero (1989).
B. Drawing from Martof (1953) illustrating the numbering system for toe clipping anurans. See text for details.
C. Labeling system for petadactylous tetrapods by Waichman (1992).

TABLE 1. Alphanumeric code for toe clipping amphibians and reptiles from Waichman (1992). Codes A5 and B5 are available only for animals with five foretoes.

ONE	A2A5	A5B5	B4C1	C5D3	A1A4B2	A2A4B5
TOE	A2B1	A5C1	B4C2	C5D4	A1A4B3	A2A4C1
	A2B2	A5C2	B4C3	C5D5	A1A4B4	A2A4C2
A1	A2B3	A5C3	B4C4	D1D2	A1A4B5	A2A4C3
A2	A2B4	A5C4	B4C5	D1D3	A1A4C1	A2A4C4
A3	A2B5	A5C5	B4D1	D1D4	A1A4C2	A2A4C5
A4	A2C1	A5D1	B4D2	D1D5	A1A4C3	A2A4D1
A5	A2C2	A5D2	B4D3	D2D3	A1A4C4	A2A4D2
B1	A2C3	A5D3	B4D4	D2D4	A1A4C5	A2A4D3
B2	A2C4	A5D3	B4D5	D2D5	A1A4D1	A2A4D4
B3	A2C5	A5D4	B5C1	D3D4	A1A4D2	A2A4D5
B4	A2D1	A5D5	B5C2	D3D5	A1A4D3	A2A5B1
B5	A2D2	B1B2	B5C2	D4D5	A1A4D4	A2A5B2
C1	A2D3	B1B3	B5C3		A1A4D5	A2A5B3
C2	A2D4	B1B4	B5C4	THREE	A1A5B1	A2A5B4
C3	A2D5	B1B5	B5C5	TOES	A1A5B2	A2A5B5
C4	A3A4	B1C1	B5D1		A1A5B3	A2A5C1
C5	A3A5	B1C2	B5D2	A1A2B1	A1A5B4	A2A5C2
D1	A3C1	B1C3	B5D3	A1A2B2	A1A5B5	A2A5C3
D2	A3C2	B1C4	B5D4	A1A2B3	A1A5C1	A2A5C4
D3	A3C3	B1C5	B5D5	A1A2B4	A1A5C2	A2A5C5
D4	A3C4	B1D1	C1C2	A1A2B5	A1A5C3	A2A5D1
D5	A3C5	B1D2	C1C3	A1A2C1	A1A5C4	A2A5D2
	A3D1	B1D3	C1C4	A1A2C2	A1A5C5	A2A5D3
TWO	A3D2	B1D4	C1C5	A1A2C3	A1A5D1	A2A5D4
TOES	A3D3	B1D5	C1D1	A1A2C4	A1A5D2	A2A5D5
	A3D4	B2B3	C1D2	A1A2C5	A1A5D3	A3A4B1
A1A2	A3D5	B2B3	C1D3	A1A2D1	A1A5D4	A3A4B2
A1A3	A4A5	B2B4	C1D4	A1A2D2	A1A5D5	A3A4B3
A1A4	A4B1	B2B5	C1D5	A1A2D3	A2A3B1	A3A4B4
A1A5	A4B2	B2D1	C2C3	A1A2D4	A2A3B2	A3A4B5
A1B1	A4B3	B2D2	C2C4	A1A2D5	A2A3B3	A3A4C1
A1B2	A4B4	B2D3	C2C5	A1A3B1	A2A3B4	A3A4C2
A1B3	A4B5	B2D4	C3C4	A1A3B2	A2A3B5	A3A4C3
A1B4	A4C1	B2D5	C3C5	A1A3B3	A2A3C1	A3A4C4
A1B5	A4C2	B3B4	C3D1	A1A3B4	A2A3C2	A3A4C5
A1C1	A4C3	B3B5	C3D2	A1A3B5	A2A3C3	A3A4D1
A1C2	A4C4	B3C1	C3D3	A1A3C1	A2A3C4	A3A4D2
A1C3	A4C5	B3C2	C3D4	A1A3C2	A2A3C5	A3A4D3
A1C4	A4D1	B3C3	C3D5	A1A3C3	A2A3D1	A3A4D4
A1C5	A4D2	B3C4	C4C5	A1A3C4	A2A3D2	A3A4D5
A1D1	A4D3	B3C5	C4D1	A1A3C5	A2A3D3	A3A5B1
A1D2	A4D4	B3D1	C4D2	A1A3D1	A2A3D4	A3A5B2
A1D3	A4D5	B3D2	C4D3	A1A3D2	A2A3D5	A3A5B3
A1D4	A5B1	B3D3	C4D4	A1A3D3	A2A4B1	A3A5B4
A1D5	A5B2	B3D4	C4D5	A1A3D4	A2A4B2	A3A5B5
A2A3	A5B3	B3D5	C5D1	A1A3D5	A2A4B3	A3A5C1
A2A4	A5B4	B4B5	C5D2	A1A4B1	A2A4B4	and so on

smallest toes in their study of *Atelopus oxyrhynchus*. Dole (1965) did not find any increase in mortality due to toe clipping in a two-year study of *Rana pipiens*. The most serious criticism of toe clipping anurans was raised by Clarke (1972), reporting that the probability of recapture of *Bufo fowleri* decreases as the number of digits excised increases. His technique involved the removal of 1 or 2 toes per foot, complete removal on the front feet, but only to the webbing on the back. Another report by Daugherty (1976) indicates a problem with weight loss in *Rana pipiens* which were toe clipped. Halliday (1994) raised the issue of concern about toe clipping anurans not only from the detrimental effect it might cause, but also on ethical issues related to using invasive or mutilation techniques. In 1995, Reaser responded with comments suggesting that very little good data exists on the actual impact of toe clips on frogs and that more standardized analysis of its use should be undertaken. She also reviews some alternative techniques that might be useful for certain species, but defends the use of toe clipping when necessary as information on populations is so important to conservation efforts. Halliday (1995) follows up with some personal communications with scientists indicating that studies of *Atelopus elegans, A. carbonerensis, Bufo marinus, B. granulosus* and *B. bufo* have little or no evidence for any impact on survival from toe clipping. Assessment and concern about use of toe clipping of *Hyla labialis* is reported by Luddecke and Amezquita (1999). The impact of toe clipping on *Rana pretiosa* was reported by Reaser and Dexter (1996) to be slight in that fewer than one percent showed any sign of infection and no mortality was found. They stress that hygienic, sterile techniques should be used and impacts carefully monitored in all toe clip studies. The implications of these reports are far reaching for the use of toe clipping with all amphibians and reptiles. Even considering these disadvantages, toe clipping is still the most common marking technique used for anurans.

ETHICAL ISSUES RELATED TO TOE CLIPPING AMPHIBIANS

In recent years there has been increasing concern over the possible impact of various marking techniques on animals. Mutilation techniques such as toe clipping in amphibians have been reviewed in several studies in order to determine the impact on survival. In addition ethical issues relative to the possibility of inflicting pain on the organisms are of concern. University Institutional Animal Care Committees (IACUC) are more inclined to request documentation on marking techniques and consideration of several alternatives.

Parris and McCarthy (2001) report evidence that toe clipping of anurans decreases their rate of return in mark-recapture studies. They re-analyzed data from four studies of *Bufo fowleri, Crinia signfifera* and *Hyla labialis* with "proper attention to statistical power" and found in three of the four studies a significant decline (6-18%) in the probability of recapture for each toe clipped after the first. Therefore this study would encourage the use of as few toes as possible if they are being clipped as a marking technique and that researchers control for the impact of clipping on the rate of recapture.

Similarly, McCarthy and Parris (2004) re-analyzed data from anuran studies employing toe clipping using Bayesian statistics and found a reduction in recapture rate of from 4 to 11% for each toe removed after the first. When the authors considered the cumulative impact of toe clipping, the recapture rate with two toes clipped was estimated to be 96% of those with only one removed. They found this ratio to decrease to about 28% if a total of eight toes were clipped. Having fairly consistent results among studies, McCarthy and Parris (2004) recommend caution when estimating population size and survival rates if toe clipping is used for marking. They also point out that these results have "important implications for the ethical treatment of animals, the continued use of toe clipping to mark species of conservation concern, and the removal of multiple toes from an individual frog or toad".

This debate continues with regards to the use of toe clipping in anurans as exemplified by a recent exchange of ideas in the journal *Nature*. May (2004) reviewed the articles mentioned in the above two paragraphs and brought the issues of both the ethics of certain marking techniques and the impact of the techniques on the data to a broader audience. In response to May's view, Funk et al. (2005) point out that some studies have found no effect with toe clipping and alternative techniques may not always be available. They point out that other invasive techniques may also have a negative impact on organisms. Funk et al. stress that the actions of IACUCs and the careful consideration of all available marking alternatives by biologists are assuring good choices. They also point out that toe clippings are now routinely used as sources of DNA for genetic study and disease and toxin identification associated with amphibian declines. They conclude with their belief that "it is less ethical to sit back and watch species slip into extinction than it is to use the best available methods to help to conserve them".

BRANDING

Kaplan (1958) described a branding technique for frogs which involves the incorporation of India ink into scarified skin of the venter (Figure 2A). Numeralized grooves are etched in the skin with a hypodermic needle and then filled with India ink. Erythema disappeared quickly, no infection was observed and the numerals remained distinguishable for over three months. Kaplan (1958) reported on an electric tattooing technique as an improvement upon the scarification method. He used a small electric tattoo marker with a narrow needle to write fine numerals or letters on the ventral surface of frogs. Higgins India ink was found most effective and was mixed with a drop of glycerin to aid its spreading into the skin. Surplus ink was wiped away, leaving a clear, permanent mark. Some initial inflammation was observed in the frogs, but was only rarely prolonged.

Heated branding irons have been used to mark toads and frogs (Clark, 1971). The irons were shaped into numerals from a wire which was 20% chromium and 80% nickel, B and S gauge No. 24 (Hoskins Chromel "A" resistance wire made by The Malin Co., Cleveland, Ohio, USA). This wire proved best because it can be heated red-hot repeatedly without oxidizing appreciably. The brands were heated with a propane torch or Bunsen burner and applied to the ventral side long enough to penetrate the dermal layer. Clark reported that recaptured individuals of *Bufo valliceps* "showed a glossy, brown, horny layer over the area of each numeral." In most toads, this layer disappeared within two weeks. Legible brands were recorded on toads for up to 21 months. This technique was also used successfully with *Rana catesbeiana* and *R. pipiens*; data from these species suggested that the branding method is similar to toe clipping in its effect on survival.

Branding of anurans has also been accomplished with the use of silver nitrate (Thomas, 1975). Seven adult *Hyla cinerea* were branded using commercial silver nitrate applicators (75% silver nitrate and 25% potassium nitrate) of the type often used for veterinary cauterization. Narrow lines were traced from one to three times on the back of the frog causing a brown mark to form immediately. One individual accidentally had the mark smeared over most of its back and died within five days. The other tree frogs were apparently not adversely affected by the marks. Within about two weeks, the dark mark fades into a light mark which also contrasted well with the background color of the frogs. Thomas recommends this method for dark colored amphibians and found the marks to remain distinguishable for over 9 weeks. The silver nitrate applicator needs to be moistened frequently when marking dry-skinned amphibians.

The silver nitrate technique seems inappropriate for marking large numbers of anurans for individual recognition. It does, however, appear to be a good alternative to some other methods in that the mark is on the dorsal surface and recapture may not be necessary for identification.

Daugherty (1976) has freeze branded *Ascaphus truei* with branding irons fashioned from No 12 A. W. G. insulated copper electrical wire. The numerals which were shaped at the insulation–stripped end of the 12 cm irons were cooled in chipped dry ice for a minimum of 30 minutes prior to initial use; continued use required only about 30 to 60 seconds of reimmersion.

Figure 2. A. Kaplan's (1958) illustration of numerals on the venter of an anuran which incorporate India ink into scarified tissue.
B. A numbered bird band used to tag an anuran toe (redrawn from Kaplan, 1958).

Brands were made on the ventral surfaces of the frogs for about 10 seconds and there was no penetration of the integument by over application. The freeze brands could be easily read within one day and brands were observed to last more than two years in the field. The mark gradually lost pigment, so that after one year the viscera were often visible through the integument. There was some indication (from one frog) that brands may not remain visible on individuals marked just after metamorphosis. All other recaptured adults and juveniles (N = 98) had clearly visible brands remaining. The author indicates that this technique has a minimal impact on an anuran's life history in that the brand is inconspicuous and produces "no continuing energy drain on an individual". An important consideration in the field use of freeze branding is the availability and cost of dry ice.

A pressurized fluorescent marking technique was used on *Eleutherodactylus podiciferus*, a small leaf litter frog in Costa Rica, by Schlaepfer (1998). A yellow, fluorescent granular powder was applied to the hindlegs of the frogs using pressurized air from a SCUBA tank. Schlaepfer (1998) compared this technique to toe clipping and found it to be more difficult, more harmful, less convenient and more expensive that toe clipping for these frogs.

Wisniewski et al. (1980) used a "Panjet" inoculator to mark *Bufo bufo* in a breeding and migratory movement study. The apparatus (obtained from F. W. Wright, Dental Manufacturing Co., Kingsway West Dundee, Scotland) uses a glass reservoir for a dye solution of Alcian-Blue which is propelled from the instrument by a mechanical plunger. The inoculator was held 5 mm or less from the undersurface of the toad limbs at a 45-degree angle for marking. Eight marking sites were used on the four limbs, one each above and below the elbow and knee, giving a variety of combinations for specific identification marks. The authors reported few injuries from this technique and marks that were easily seen and some were retained over a period of at least one year. Disadvantage noted were that the marks are small so were best read by the initial investigators doing the inoculations. In addition, toads were sometimes difficult to hold while administering the mark and if any of the marks failed over time the code combination accuracy for identification was reduced.

TAGGING AND BANDING

Woodbury's (1956) review of the jaw tagging technique of Raney (1940) did not include the report by Stille (1950) that the loss of these tags by toads was significant. Woodbury (1956) did cite a personal communication with Stebbins stating that the jaw tags were often lost and "caused considerable irritation". Raney and Lachner (1947) studied the impact of these jaw tags on growth in *Bufo americanus* and reported inconclusive results, but a negative impact was indicated. This technique is no longer used.

Kaplan (1958) used aluminum toe bands to tag frogs (butt-end bird band, #1242, size 2 , National Band and Tag Co., Newport, KY 41071). These numbered, cylindrical bands were placed around a toe (see Figure 2 B) and the two ends were passed together with pliers. The bands were tightened so as not to restrict circulation, but they pierced the webbing of the foot. These tags remained fixed indefinitely and caused no apparent reduction in the movement of the frog.

A glass bead tag is described by Nace and Manders (1982) for individual *Xenopus laevis* in the laboratory where specimens were anesthetized with MS-222 (tricane methanesulfonate) after being cleaned and rinsed. The left forelimb, lateral to the humerus (or hindlimb medial to the femur) was pierced by at 21 gauge hypodermic needle through which a 00 (28 gauge) surgical Steel Monofilament Type B non-Capillary (or 32 gauge Ethicon Sutupak® surgical wire) was passed and then the needle withdrawn. After stringing a sequence of up to 4 colored glass beads (allowing 9999 combinations) on the wire, "a loose loop was made and secured by a square knot tied on the medial (or lateral) side of the limb, and the wire leads were clipped". Nace and Manders (1982) found healing was good without using antibiotics or fungicides and tag retention for up to three years in the lab, but the technique is not recommended for field use due to possible snagging on substrates.

Watson, et al. (2003) used marked *Rana pretiosa* in their second active season (> 42 mm SVL) with numeric-coded fingerling tags which were attached over the knee with elastic thread following the technique developed by Elmberg (1989). McAllister et al. (2004) recommended against use of such knee tags in *R. pretiosa* due to skin and muscle lacerations in 33% of the recaptured specimens. Robertson (1984) marked small *Uperoleia rugosa* with a unique combination of Scotchlite (3M Company brand name) reflective sheeting cut into 1 mm X 1 mm squares. After wiping dry the skin surface of the frog, Robertson attached the squares with fast-setting cyanoacrylate tissue cement which would keep them in place for up to ten days. By inhibiting the secretions of skin glands under the patches with a freeze brand, the squares remained fixed for from three to six weeks. Robertson used toe clipping as a back up and more permanent mark in case the more easily seen reflective squares were lost.

The use of a radioactive tag (CO^{60}) in a study of *Bufo canorus* is described in detail by Karlstrom (1957). With the development of better telemetry techniques and the use of PIT tags, radioactive tagging is not currently a tag of preference. Logistics and environmental concerns also make the use of radioactive isotopes in the field less desirable.

A nylon waistband was used by Emlen (1968) to aid in the behavioral study of male *Rana catesbeiana*. This tag was chosen because it allowed individual identification in the field without disturbing the bullfrog. Waistbands were made from 13 mm wide, preshrunk, nylon elasticized banding. These were about 13 cm long and painted with distinctive, colored patterns or black numerals using Testor's butyrate dope paint. Each strip was then stapled into a circular band looped over the bullfrog's hind legs and positioned around the pelvic girdle. The bands were fit snugly so as to prevent rotation or catching on items in the environment, but loosely enough to permit normal activity and respiration. Color-marked waistbands were recognizable from maximum distances of about 8 to 12 m using a headlamp and binoculars, while numbered bands were only visible from 4 to 6 m. Emlen reported no differences in behavior, mortality, emigration rates, or weight loss between tagged and untagged frogs. Due to soiling and staining, seasonal replacement of these waistbands would be necessary.

Robertson (1984) reported a negative impact of Emlen's (1968) waistband design on ten species of Australian hylids and leptodactylids. A modified waistband consisting of surgical thread sewn through a square piece of surveyor's flagging worked well on *Rana clamitans* (Rice and Taylor, 1993). This thread was then tied around the waist of the frogs so that the flagging patch rests flat on the dorsal surface.

Fluorescent yarn tags were used to track the movement of *Rana catesbieana* and *R. clamitans* by Windmiller (1996). Two to four, 100-140 mm lengths of acrylic yarn were arranged in parallel. These were then glued to a 15 mm square of self-adhering veterinary bandage at one end using a few drops of heat-melted glue. A second bandage square was then glued onto the first, covering the yarn ends. After patting dry the mid –dorsum of the frog, a drop of chanoacrylate glue was place on one side of the tag assemblage and pressed onto the frog's dry integument for one minute. Once fastened to the dorsum of the frog, the tag was then suspended in a plastic bag filled with fluorescent pigment (Radiant Color, Richmond California, USA) and the pigment pressed into the yarn. The pigment trails of released frogs were followed the next night by illuminating them with a long-wave (366 nm) ultra-violet lamp (Blak-Ray™ Model ML-49, UVP, Inc., San Gabriel, California) while wearing safety/UV enhancing glasses. These tags only remained on frogs from one to two days in the field so toe clipping was used for long term individual recognition. Tags were easily removed with gentle pressure and no mortalities were noted in 55 trials. This technique allowed the researcher to test the relationship of the path orientation of the frog to such environmental variables as substrate and topography.

Prefabricated, biocompatible fluorescent tags were tested on five species of anurans by Buchan et al. (2005). These Soft VIAlpha Numeric Tags (Northwest Marine Technology, Inc., Shaw Island, Washington) were each imprinted with a letter (A-Z) and a number (00-99) and are available in several color combinations (e.g., black letters on yellow orange or red fluorescent backgrounds). Smaller (1.0 X 2.5 mm) or larger (1.5 X 3.5 mm) tags were used depending on the size of the specimen. A forceps or injector (Northwest Marine Technology) was used to insert the tag into an incision that was made by a standard blood lancet in the subcutaneous layer near the sartorius muscle on the ventral side of the leg. Buchan et al. (2005) found high survivorship, retention and readability of these tags in both the laboratory and the field. Somewhat better results were found when the tags (US $1.00 per tag) were inserted in the incision with the injector (US $120.00).

Using *Rana sylvatica*, Moosman and Moosman (2006) studied how the placement of Visible Implant Elastomers (VIE) was related to subcutaneous movement of the marks and attempted to identify the optimal regions for implant placement. They found that over one month's time marks on the limbs and dorsal side showed less migration than ventral and mid-body marks. It was also found that handling the specimens soon after injection of the elastomer can result in mark movement and that skin pigmentation can partially obscure the color of the mark or the mark itself. Fifteen of a total of 48 marks used in this laboratory study had subcutaneous movement.

Windmiller (1996) used luminescent capsules filled with Cyalume™ to both observe the movement of anuran species from a distance and to then document events such as the influence of conspecifics on their path choice. Cyalume™ fluid (American Cyanamid Co., Wayne, New Jersey, USA) was placed in Eppendorf micro-centrifuge tubes so they were half full and then securely capped. These capsules were then marked for individual identification and glued onto adhesive bandage squares. These markers were then glued to the frogs dorsum as described above for fluorescent yarn tags. Smaller and lighter weight Eppendorf capsules than the 0.5 ml or 1.5 ml sizes can be customized by cutting them down in size and resealing the pointed tip with paraffin. Windmiller warns that Cyalume™ fluid is fatal to *Rana catesbeiana* and likely is to other anurans so contact must be avoided. Different colors of the fluid were used to distinguish different release groups of frogs that were then observed at night for up to two hours and from distances up to 80 m. About 10% of the frogs lost their capsules, fewer if they were affixed in the lab rather than the field. The tags could be removed with gentle pressure.

TRAILING DEVICE

A trailing device using nylon sewing thread was devised by Dole (1965) to study the movements of individual *Rana pipiens*. A sewing machine bobbin was placed in a holder constructed from a piece of rigid plastic tubing with a flat plastic base glued to one side and then attached on a 6.3 mm wide, elastic band. The bobbin was held in place with a wire passing through two holes in the top of the holder. The thread trailed out through a slot cut in the back of the holder and the entire device was secured around the waist of the frog. A small stake was used to mark the point of capture and the trailing thread was tied to it. The 50 m of thread on the bobbin lasted from one hour to seven days, depending on the activity of an individual. Empty bobbins were easily replaced in the field without removing the elastic saddle. The loaded trailing device weighed about 8.5 g and "did not seem to hinder the animal's jumping ability seriously." The device may have shortened their jumping length and no individuals less than 60 mm long were trailed. Dole reported some difficulty with frogs swimming and entering crevices with the device. If waistband irritation occurred on the skin, as it occasionally did, the device was removed immediately until complete healing took place. Grubb (1970) modified Dole's technique by using 200 m of cotton thread while studying *Bufo valliceps..*

The spool-and-line tracking device of Wilson (1994) was adapted for use with *Leptodactylus labyrinthicus* by Tozetti and Toledo (2005) using quilting cocoons (Spiltex, Ltd.) holding 300 m of cotton thread. A food wrap plastic (e.g. Saran Wrap®) was used as a case for the thread cocoon with a piece of 2 cm wide adhesive tape wrapped around it. The thread was pulled through a small hole made in the cover and then the package was attached with a 1 cm wide elastic band belt around the inguinal region. The device was removed after a maximum of three days and the 8 g weight of the trailing device never exceeded 5% of the frog's body weight. They authors report that the device seems to have little or no impact on the activity of large bodied anurans (Tozetti and Toledo, 2005).

Another tracking method using fluorescent powder pigments (Radient Color, Richmond, California, USA) with recently metamorphosed *Rana sylvatica* has been reported by Rittenhouse et al. (2006). They monitored short-term (water loss) and long-term (survival and growth) effects of the marked (covered in pigments of varying colors) individuals and no impact was found. The authors suggest that this technique might be preferred over radio-telemetry and thread-trailing is some uses due to its low cost, providing a detailed record of the movement path, and possible use with small individuals. A possible disadvantage to the technique could be increasing the visibility of the frogs to predators.

PATTERN MAPPING

While more commonly used in salamanders, the patterns of amphibians may be unique enough to make individual recognition possible. Kurashina et al. (2003) found individuals of *Rana porosa brevipoda* have distinctive dot patterns on the body that can be used to identify individuals. Since these spot patterns do change with growth, this photographic technique is only useful for short-term studies of froglets. Wengert and Gabrial (2006) used photographs of chin spot patterns in *Rana mucosa* and found a high success rate in reidentification over a three month study; changes in spot pattern over time were reported so that regular recaptures might be needed to minimize misidentifications. The effectiveness of using photographic identification on long-term studies of adult anurans has yet to be determined.

Pointing out concerns in using artificial or invasive marks with the Nationally Critical and endemic *Leiopelma archeyi* (Archey's frog) in New Zealand, Bradfield (2004) developed a highly accurate photographic identification technique using natural markings. This study is a model on how to develop and test a photographic pattern mapping technique. A series of six digital photographs were taken from dorsal, ventral, cranial (facial), caudal, right lateral and left lateral views. Photos using a flash were most successful in discerning usable characteristics if the black markings below the eye and black upper lip markings were continuous or discontinuous on the right and/or left sides. A filing system for the photos used subgroupings which substantially reduced the amount of time needed to make identification of recaptures. Intra- and inter-observer consistency of identification was tested and found to be high. Overall, 99.2 % of identifications were successful once recaptures were assigned to their subgroups.

SALAMANDERS

MUTILATION TECHNIQUES

At the time of Woodbury's (1956) review, toe clipping was the only successful marking technique reported for salamanders and was just beginning to come into use. Toe clipping has been the most common means of marking salamanders in recent years and in most cases something similar to Martof's (1953) system (see Figure 1) of marking anurans is used. The system used by Twitty (1966) in marking *Taricha* is one example (Figure 3). Another is the one discussed earlier by Waichman (1992) as shown in Figure 1 C and Table 1.

Table 2 summarizes some reports in the literature concerning the regeneration of digits in salamanders. It appears, then, that regeneration needs to be considered, but whether or not it is a deterrent to using toe clipping depends upon the species. Heatwole (1961) controlled regeneration of digits in *Plethodon cinereus* with a beryllium nitrate treatment. He found concentrations of beryllium nitrate 0.1 N or greater completely inhibited regeneration ($N = 6$) and recommended using the lowest possible dosage to avoid any potential harmful effect. Heatwole simply dipped the limb in the beryllium nitrate solution for 1.5 minutes and then rinsed it with distilled water. Beryllium nitrate should be used only with extreme caution in that is it toxic to man and can easily diffuse through the skin. Extreme care, rubber gloves and forceps are important precautions if using beryllium nitrate. Efford and Mathias (1969) reported that neither food consumption nor exposure to beryllium nitrate solution affected regeneration in *Taricha granulosa*. They did discover, however, that temperatures below 10° C inhibited regeneration of digits, which would be an important factor in field studies. At 18° C, *T. granulosa* had a significant amount of toe regeneration.

Healy (1974) marked post-larval metamorphs of *Notophthalmus v. viridescens* by "amputating one limb at the middle of the zeugopodium." He questioned the influence of this on the mortality rate because relatively few individuals were recaptured. Juvenile (<2.5 cm SVL) *Desmognathus f. fuscus* have toes too small for precise clipping, but have been successfully marked by clipping small pieces of the tail at either right angles to the longitudinal axis of the body or in the transverse plane at different angles (Orser and Shine, 1972). This technique allowed groups of marked juveniles to be distinguished for at least one month before regeneration became a problem.

A unique mutilation method is that of autotransplantation which has been used for *Triturus alpestris* (Rafinski, 1977). These newts were anesthetized with 0.1% MS 222 and 3 mm square pieces of skin were removed from the orange venter and the bluish to brown-black dorsum using fine forceps and scissors. These grafts were exchanged and needed no adhesives. After transplantation, which took about 3 minutes, the newts were kept isolated out of the water for an hour. Individual recognition was possible by using various combinations of the locations and number of grafts. Rafinski reported a 95% success rate and retention of the grafts for at least 3 years. This technique may prove very useful for long term marking of many salamanders, but is more difficult to use with species having spotted patterns and cannot be used with larvae.

Peterman and Semlitsch (2006) report on the importance of properly buffering solutions of MS-222 (tricaine methanesulfonate) and the effect of various concentrations on the time to become anesthetized and recover in four plethodontid salamanders. They found that induction time usually decreased and the time needed for recovery increased with increasing concentrations of MS-222. It is recommended that the results of this study be consulted before using this anesthetic with salamander marking techniques or other surgeries.

No significant short-term effects of toe clipping on survival or growth were found in a study of *Ambystoma opacum* (Ott and Scott, 1999). No more than two toes were clipped on each animal and some regeneration was observed. However, cutting two adjacent toes on a foot seemed to be more recognizable for two to three years than clipping one to two toes on different feet due to regeneration and accidental loss of toes. Ott and Scott (1999) concluded that toe clipping for *A. opacum* and perhaps other species of salamanders is a more economical and practical method over PIT-tags and pattern mapping.

In a similar study, Davis and Ovaska (2001) found that toe clipping in *Plethodon vehiculum* resulted in less weight gain than those marked with fluorescent tags or used as controls. They used unique combinations of three clipped toes per salamander, removing no more than one per foot. The number of ambiguous marks in toe clipped individuals increased dramatically

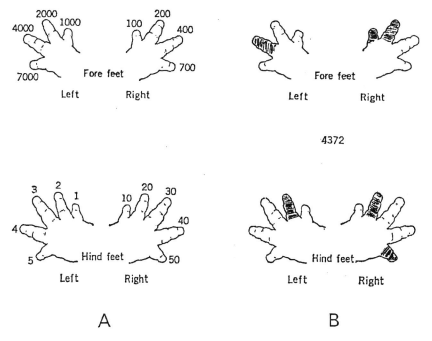

Figure 3. The numbering system used by Twitty (1966) in toe clipping salamanders.
    A.   The numerals assigned to each digit
    B.   In order to obtain an individual numbered 4372, the shaded toes would be clipped.

in frequency over the fluorescent tagged after 50 days, but the majority of both retained useful marks throughout the 87-week field study.

Arntzen et al. (1999) toe clipped and tail clipped *Triturus cristatus* in comparing the impact of these techniques finding that tail clips regenerated more rapidly than toe clips. They found no significant difference in survival of those toe clipped newts with and those without tail clip. Due to faster regrowth found in the tail clipped newts, the authors recommend using tail clips when doing tissue sampling and toe clipping for population studies.

In a study of the impact of invasive marking techniques on *Desmognathus fuscus* and *D. monticola*, Kinkead et al. (2006) concluded that toe clipping done with no anesthesia is a humane and successful technique. They measured stress hormone responses (adrenaline and noradrenaline levels) in elastomer injected, toe clipped without anesthesia, toe clipped with local and whole body anesthesia as well as controls (handled, but not marked) and found no significant difference among the trials or on the behavior of the salamanders.

BRANDING

Taber, et al. (1975) branded *Cryptobranchus alleganiensis* with heated 1.5 cm high, numerical brands of thin nichrome wire. In most cases, these brands remained clear through a two year study, but some did fade and required rebranding. This technique is similar to that Clark (1971) used on anurans and could probably be modified for caudates. The freeze branding technique of Daugherty (1976) was not used with salamanders, but also seems applicable. In 1983, Bull et al. tested freeze branding application time in marking *Ambystoma macrodactylum*. Five mm high letter brands made with copper rods having silver tips were immersed five minutes in liquid nitrogen They found that a 0.75 second application produced the most clarity with maximum clarity occurring at three months. Woolley (1962) used a black, Carter's felt ink pencil to mark salamanders. Dilute acetic acid or ammonium hydroxide was used to remove slime on the salamander's tail before application. These marks lasted at least a month.

Taylor and Deegan (1982) inject a fluorescent pigment into *Notopthalmus viridescens* dermis using a compressed air spray gun resulting in minimal mortality. After 39 days individuals began succumbing to *Saprolegnia* infections, but these could not be linked to the use of fluorescent branding.

A technique using dry fluorescent dust applied with pressurized air on *Plethodon jordoni* and *P. glutinosus* was described by Nishikawa and Service (1988). Materials needed for this technique include: four to 10 colors of inert fluorescent pigments (Scientific Marking Materials, Inc., P.O. Box 23122, Seattle, WA, 98124, USA or Radiant Color Co., 2800 Radiant Ave., Richmond, CA 94804, USA), canisters, spray gun with 0.25 inch nozzle, hose, single-stage regulator and pressurized air. Specimens were placed on a dry enamel pan and the spray gun held 1 cm away from the skin surface using lower pressures (25 psi) for smaller specimens and greater pressures (40 psi) for larger ones depending too on the size of the fluorescent particles used.

Table 2. Digit Regeneration Times for Selected Toe Clipped Salamander Species.

Species	Time Interval for Regeneration	Source
Taricha spp.	Several years	Twitty (1966)
Taricha granulose	Indefinitely, when kept below or at 10° C	Efford and Mathias (1969)
Plethodon cinereus	7 months in laboratory	Heatwole (1961)
Plethodon glutinosus	At least two years	Wells and Wells (1976)
Plethodon wehrlei	50% after 100 days; regenerated digits identifiable by lack of pigmentation	Hall and Stafford (1972)
Cryptobranchus alleganiensis	One year	Hillis and Bellis (1971)
Batrochoseps spp.	Slow and regenerated toes recognizable	Hendrickson (1954)

Three to four marks were generally all that could be easily applied to each animal on either side of the body (cranial and caudal to the forelimb, midbody, cranial and caudal to the hindlinb) for a maximum of ten possible locations. Nishikawa and Service (1988) reported higher recapture rates than for any salamander toe clipping studies done to that time.

TAGGING AND TRAILING

  Floy T-tags were used successfully to mark *Cryptobranchus alleganiensis bishopi* by Nickerson and Mays (1973), but no details of the procedure were discussed. Whiteford and Massey (1970) tagged *Ambystoma tigrinum* by suturing a numbered, plastic float through the tail with a monofilament line. The line was made long enough to allow these salamanders to move through the deepest portion of the lake.

  Woolley (1973) tagged *Eurycea lucifuga* and *E. longicauda melanopleura* with a subcutaneous injection of two parts Liquitex Acrylic Polymer to one part distilled water. This mixture was injected into the lateral proximal caudal region with a 22 gauge hypodermic needle, leaving a mark 7 to 10 mm in diameter. Woolley observed no adverse effects with this procedure and found slight fading in only a very few individuals. Although this study did not require the recognition of individuals, the author suggests that a series of acrylic colors could be used to differentiate individuals. Woolley found this to be a convenient, relatively harmless technique allowing quick recognition in the field with a reduction in subsequent handling.

Subcutaneous injections of fluorescent, elastomer-dyes were used in *Plethodon vehiculum* by Davis and Ovaska (2001). Three dye colors (red, orange and yellow Visual Implant Elastomers, VIE, from Northwest Marine Technology, Inc., P.O. Box 427, 976 Ben Nevis Road, Shaw Island, WA 98286 USA) were injected into six location in various combinations to create 816 unique marks. The elastomer was mixed with a hardener following the manufacturer's instructions and 0.1 cc was placed in a 0.3 cc syringe. For injection the syringe was placed in a custom made holder to facilitate manipulation and minimize risk of over-insertion of the 29-gauge 12.7 mm needle (see Figure 4). The authors recommend fluorescent marking or pattern mapping over toe clipping in studies of plethodontid salamanders.

Bailey (2004) evaluated Visual Implant Elastomer (VIE) marking in *Eurycea bislineata wilderae* by documenting mark retention, salamander survivorship and growth rates. The VIE and viewing lights were provided by Northwest Technologies, Inc. Salamander were placed in plastic zip-log bags, cooled down for 5 to 10 minutes and then injected through the bag with a 0.3 cc insulin syringe. One to three colors were placed in two of four possible locations in each individual (anterior to either hindleg or posterior to either foreleg). Prior to injections needles were treated with 95% EtOH. No adverse effects and no loss of marks was found over the 11 month study. Misreading of marks in the field by observers was found to be a factor in some cases, so a training period for investigators is recommended.

Rittenhouse et al. (2006) used the same technique with powdered fluorescent pigments as described for *Rana sylvatica* above with newly metamorphosed *Ambystoma maculatum*. No short or long-term effects of the powder dusting on the salamanders were detected.

RADIOACTIVE TAGGING

As mentioned in the introduction, radioactive tagging is not used as much as it was several decades ago and licensing and environmental issues need to be carefully considered. Barbour, et al. (1969) tagged *Desmognathus fuscus* with a 0.75 x 3.3 mm piece of 45 ìc ^{60}Co alloy wire. These wires were placed in the tail musculature with a modified hypodermic needle. No adverse effects were observed. A detection apparatus (Thyac II model 489 survey meter) located tagged individuals at a distance of from 3 to 4 m. when they were at or near the surface. *Plethodon jordani* have been tagged with similar injections of 18 gauge, 20-48 ìc, ^{182}Ta wire (3 to 5 mm long) into the abdomen for orientation, homing and home range study (Madison and Shoop, 1970).

Ashton (1975) used ^{60}Co (35-50 ìc) alloy wire (ICW Pharmaceutical, C & R Division, 2727 Campus Dr., Irvine, CA 92715 USA), about 3.0 x 0.5 mm in size to tag *Desmognathus fuscus*. The tag was inserted as described by Barbour, et al. (1969) and monitored with a Victoreen Thyac III Model 491 survey meter with a scintillation probe. In this study and others with plethodontids, Ashton has observed local ulceration which eventually opened, exposing the tag. Ashton reports that this "burn out" has not been a problem with *Pseudacris, Necturus* or in reptiles. He also reports that platinum coated wires are excellent for reducing local tissue damage in small animals and that sterile procedures are important to avoid infection.

The impact of using radioactive wire tantalum-182 on *Ambystoma talpoideum* survivorship and body condition was reported by Semlitsch (1981). He concluded that these tags can be used with no apparent mortality for locating the salamanders in the field for a period of about one month. However, longer exposure to the radiation results in tag loss, weight loss and possible tissue damage.

PATTERN MAPPING

Perhaps the most ideal method of recognizing individuals is to use their own, naturally occurring variation in morphology. This has been done for *Triturus cristatus* and *T. vulgaris* by photographing belly patterns (Hagstrom, 1973). In using this technique, one must be sure that the patterns are both recognizably different and constant through time. Healy (1974 and1975) found the variation in the dorsal spot pattern of *Notophthalmus viridescens* useful in identifying individuals. This technique requires recapture and handling which may not be consistent with the animals' behavior.

Digital photographs of dorsal spot patterns were tested for their effectiveness in the identification of *Eurycea bislineata wilderae* by Bailey

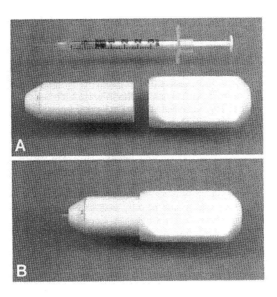

Figure 4. A. Holder and 0.3 ml insulin syringe used for fluorescent marking by Davis and Ovaska (2001).
B. Syringe inside holder.

(2004). Observers were given color printouts of the dorsal view of each animal with the SVL written below the image for a size reference. They compared these photos to the test animal and were timed as to how long it took them to identify each animal. Photo-identification rates were very high (about 94% correct), but may be lower if observers were also given un-marked (photographed) individuals in the attempted comparisons. Computer aided matching of images would likely make this technique more useful (Bailey, 2004).

Loafman (1991) photographed the natural variation of spot patterns in adult *Ambystoma maculatum* to successfully identify about 97% of the 244 individuals studied. Adult *A. opacum* were photographed in a box and successfully reidentified using their distinctive bar pattern over a one year period by Doody (1995). Due to ontogenetic changes this technique is not appropriate for juvenile *A. opacum*. Using head patterns alone, 80% of the adults could be distinguished from one another (Doody, 1995). Similarly, Lackey et al. (2006) used digital images of the ventral spot pattern variations to individually identify *Pachytriton brevipes* and found that spot patterns varied with snout-vent length which enabled easily placement into age categories.

The use of spot patterns for identification of *Ambystoma maculatum* is reviewed by Grant and Nanjappa (2006) with special attention paid to possible errors in this technique. They attempt to quantify error rates in specific mapping techniques, modify techniques to address sources of error, compare methods relative to observer bias in both the field and laboratory and determine the effort needed to search resulting databases for specific individuals. Advances in pattern recognition research will be helpful in eliminating bias and handling large samples in the future. The specific suggestions discussed by Grant and Nanjappa (2006) are recommended for researchers dealing with large samples of specimens identified by patterns.

CAECILIANS

Gower et al. (2006) review earlier publications of the use of alphanumeric fluorescent tags (VIAlpha tags) with the Indian caecilian *Gegeneophis ramaswamii*. They found that making an incision with a scalpel blade before using the injector increased the efficiency of tagging and that anesthesia was needed to quiet the specimens. Equipment was not sterilized while in use at one site, but was sterilized before switching to additional sites to minimize the risk of spreading pathogens.

TURTLES

SHELL NOTCHING

The most commonly used marking technique for turtles is that of notching the shell. Cagle (1939) developed a coding system for notching which has had wide usage (see Figure 5A). Marginal plates are numbered 1 to 11 on each side of kinosternid turtles and 1 to 12 on chelydrid, emydid and testudinid turtles. In addition, Cagle suggested notching the plastron to give a number identification to each carapace series and therefore increase the potential number of turtles that could be marked. By using four marginal notches, 2516 marking combinations could be made. Marks were recorded by having a comma separating those made on one side of the carapace, and a hyphen to separate the left and right sides. Therefore, a turtle designated 2, 6-3, 5 had the second and sixth marginals on the left side notched and the third and fifth on the right. Cagle used a triangular file to make the notch and a square-edge file to widen it to one-third to one-half the width of the plate. Young individuals with unossified carapaces were marked using a small pair of sharp scissors. Ernst (1971) used a hacksaw blade to notch adults of *Chrysemys picta* and marked juveniles with a sharp pen knife. Tinkle (1958) used Cagle's notching technique and found that marked *Graptemys* sp. and *Pseudemys* sp. were more wary than those turtle in the population which were unmarked. This change in behavior may result in decreased recaptures in mark recapture studies (Tinkle, 1958).

Ernst, et al. (1974) evaluated the notching technique as being best for marking great numbers of turtles in mark-recapture studies because of low cost and the need for few tools. However, they also proposed a new coding system (Figure 5B) which will allow the consecutive numbering of turtles as can be done with some numbering systems in the toe clipping of lizards and amphibians. A number for each individual is obtained by adding the numerical values of the marginals which have been notched. By using plastron marks on the gulars and anals, an almost unlimited number of combinations can be used. This system can be abbreviated for use with kinosternid turtles which have fewer scutes. The authors suggest that marginals at the bridge or junction of the plastron and carapace should not be notched so as not to weaken the shell.

Another modification of Cagle being used by a number of investigators is reported by Sajwaj et al. (1998). Each marginal scute is given a cranial to caudal letter designation such as A through L on the right side and N through Y on the left. One or two notches or holes are then made in the scutes and coded as the identifying mark. For example, one notch in the C and D and two in the S scutes would be read "CDSS".

Cagle (1939) cautioned that these marks may not be permanent when used in young specimens whose shells are not yet completely ossified. Therefore, he also marked juveniles by clipping the first phalanx of toes with a numbering system similar to that used for lizards.

TAGGING

Kaplan (1958) tagged turtles with numbered, aluminum bands (National Band and Tag Co., Newport, KY 41071) fastened through holes drilled in the carapace (Figure 6). *Chelydra serpentina* have been tagged with aluminum plates (4.2 x 1.6 cm) which were attached with 0.020 gauge, stainless steel trolling line through two 3 mm holes drilled through the marginal scutes just to the side of the tail (Loncke and Obbard, 1977). A letter

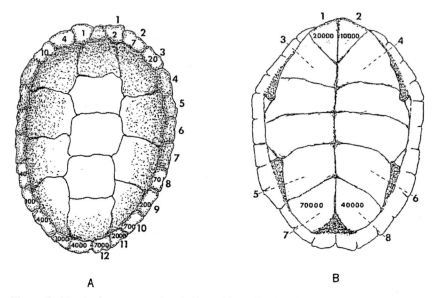

Figure 5. Numbering systems for shell notching. Cagle (1939) numbered the marginal plates on each side from 1 to 12 on the carapace (A). Plastral plates were numbered 1 through 8 (B). Ernst, et al. (1974) numbered 6 plates at the cranial end of the carapace and 10 plates at the caudal end (A). In addition they numbered four plastral scutes (B) for marking an extremely large number of turtles.

and a number were recessed into each tag and painted black to make them more visible. Loncke and Obbard evaluated their technique as being "quick, uncomplicated and successful for at least three years." They preferred it to Kaplan's method because the tag could be read at a distance reducing the need to handle marked turtles. The turtles may only need to be handled "every few years to adjust tags and allow for growth" (Loncke and Obbard, 1977).

A radioactive tag was used in a study of *Clemmys guttata* by Ward et al. (1976). A radioactive pin (tantalum^{-182}, 100 microcuries, 1 x 5 mm, half-life = 115 days) was placed in a hole (H" 0.1 mm in diameter) drilled in the left 10th or 11th marginal scute. The tag was secured with a drop of waterproof acetate glue. Marked turtles could be detected up to about 10 m away using a portable beta-gamma survey meter equipped with a scintillation probe. A similar technique had been used by Bennett et al. (1970) with *Kinosternon subrubrum*, *Pseudemys scripta* and *Deirochelys reticularia*.

Turtles have also been successfully tagged by using a Buttoneer (Dennison Manufacturing Co., Framingham, Massachusetts 01701), a small tool used to fasten buttons to clothing by means of a plastic plug (Pough, 1970). The plug comes in different sizes and has a stem with a bulb at one end and a crossbar at the other. Pough (1970) placed the plug into a 3/32-inch hole drilled through a marginal plate. He suggests that this technique is good for young turtles, where regeneration has been a problem with notching. Froese and Burghardt (1975) used the plastic buttoners to mark *Chelydra serpentina*, using a numbering system for the posterior marginal scutes. They found detachment a problem when the plugs were too short to allow free movement in the holes. However, even if the plugs were lost, the 2 mm holes were still visible. Galbraith and Brooks (1984) found the Buttoneer method inadequate for hatchling *C. serpentina* since portions of the carapace may not be adequately ossified to withstand the installation of the tag. They used plastic beads (2.5 mm in diameter and 1.5 mm thick) tied with a reef knot to the middle of a 15 cm length of monofilament nylon line or polyester thread. After placing the hatchlings in a cold torpor, both sides of the margins of the carapace were swabbed with alcohol and a marginal scute was then pierced from below with a fine-gauge hypodermic needle. With the needle in place the thread was pulled through the needle from the point side and then the needle was removed, leaving the tag with a bead on the dorsal surface and thread hanging free below. A second bead was then threaded through the tag below and the thread ends tied together. The loose ends were trimmed within 4 mm of the knot. The researchers experimented with trying to seal the knot by melting with a fine soldering pencil, but found this no more successful in retaining tags than leaving the knot unmodified (Galbraith and Brooks, 1984). The authors found a loss rate of about 2.5% per week in the laboratory and did not field test this technique.

Layfield et al. (1988) used rings of 35-lb steel trolling wire in the posterior marginal scutes of hatchling *Chelydra serpentina*. A 5 cm piece of wire was passed through a posterior marginal scute and bent to form a 3 mm diameter ring using needle-nosed pliers. Using the coding system of Cagle (1939) up to 1554 individual hatchlings can be identified. Layfield et al. (1988) recommended using colored plastic beads (Barriecraft Sales, 123 Dunlop St. East, Barrie, Ontario, Canada L4M 1A6) on the wire rings to make identification easier and increase the number of potential marks. Some minor loss of these ring tags was recorded in the lab and field so it was

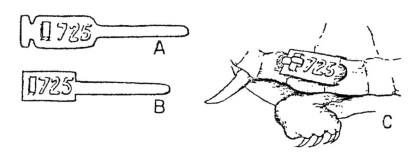

Figure 6. A. The band as it was purchased by Kaplan (1958).
B. The band as it was modified to obtain a long tab.
C. The band as it appeared when fastened through two holes in the carapace of the turtle.

recommended that they be replaced by a shell notching method when the turtles have grown to an adequate size (Layfield et al., 1988).

An inexpensive "spaghetti tag" was used to mark *Apalone spinifera* by Dreslik (1997). A 3 m length of Romex wire containing 8 different colored wires within it was first stripped of its out sheath. The 8 wires were cut into 10 cm pieces and the plastic sheath was then removed from the copper core. This provided 240 plastic tags and by using from one to four tags per animal, Dreslik created a total of 4680 unique color combinations. A small incision was made in the carapace edge and an unknotted end of a tag was threaded through the opening. The other end was also then square-knotted leaving 1.5 cm between the ends of the tag to permit growth. Excess plastic sheathing was removed. No data was provided on the longevity and loss rate of these inexpensive tags.

Testudo gigantea have been marked with numbered, titanium disks which were fixed in depressions drilled into the keratin shield (Gaymer, 1973). The tags were held in the depression by a metal-resin adhesive and had a very high retention rate. Ward, et al. (1976) marked the carapace of *Clemmys guttata* with an "adhesive" tag which bore an identification number. Graham (1986) strongly warned against the use of Peterson Disc Tags (often used on fish) as they can cause turtles to get caught in trap netting when attached in the conventional way and loosening over time and turtles may drown.

Davis and Sartor (1975) tagged or flagged turtles (*Chelydra serpentina*) by putting a 1/8 inch diameter, wooden dowel through a hole which was drilled in the nuchal scute. The dowel was held with epoxy glue and was marked at the top with colored paint or plastic tape for individual recognition. The aquatic movements of these turtles could be easily observed when they were in shallow water or near the surface. To reduce the hindrance this technique caused turtles moving in thick vegetation, two shorter dowels were connected with a piece of rubber tubing.

Sea turtles are usually marked with numbered, Monel® tags which are attached to the trailing edge of a fore flipper (for example, *Caretta caretta*: Le Buff and Beatty, 1971; *Chelonia mydas*: Carr, 1967; Pritchard, 1976; *Dermochelys coriacea*: Bacon, 1973; and *Lepidochelys olivacea*: Pritchard, 1976). Pritchard (1976) reported significant tag loss through both metal corrosion and tissue necrosis, particularly with *Dermochelys coriacea*. In an extensive study of *Caretta caretta*, Henwood (1986) concluded that "the Monel flipper tag is not a reliable permanent tag for loggerhead turtles". Henwood speculates that significant tag loss is due to factors including tag corrosion, tissue necrosis and improper tag application. Balazs (1985) found that "681 flipper tags" manufactured from Inconel® used on hatchling and adult green turtles had no corrosion and were superior to Monel® in that regard. At the time of his study Inconel® tags were difficult to obtain due to high production costs, but highly corrosion resistant titanium flipper tags (Stockbrand Co. Pty. Ltd., Mt. Hawthorn, Western Australia) were recommended as a viable option. Alvarado et al. (1993) compared tag retention rates in metal and plastic varieties which were applied to different flippers in the black sea turtle (*Chelonia agassizi*). Monel® No. 19 (National Band and Tag Co., Newport, KY, USA) and plastic (41 mm x 35 mm, Allflex

Tag. Co., USA) tags were each placed on randomly selected flippers for 131 nesting females. Their results showed that plastic tags were retained at a significantly higher rate than the metal tags over a four-year period regardless of which flipper was tagged. However, some problems with abrasion and algal growth obliterating embossed numbers on the plastic tags was noted (Alvarado et al., 1993). Medway (1978) reports that "Dalton tags", primarily designed for agricultural use, have had similar success to those discussed above. The probability of tag loss in *Chelonia mydas* was reported by Bjorndal et al. (1996) where in a small sample size Inconel® tags were not better retained than Monel® tags. These authors suggest "at least two tags should be applied to each turtle to improve long-term recognition of individuals and to allow corrections for tag loss to be calculated" (Bjorndal et al., 1996).

A study using tags on *Dermochelys coriacea* by Eckert and Eckert (1989) concluded that: 1) Monel® tags secured to foreflippers can carve furrows in the carapace due to rubbing during nesting and swimming; 2) mounting the tags cinch-tab up (up-side-down) eliminates this carapace damage; 3) plastic tags (Riese "Flexible Jumbo") give poor recovery rates, but inflict no carapace damage, and 4) tag loss "is rarely associated with an external pigment or mutilation scar". Plastic tags have the advantage of being non-corrosive (Eckert and Eckert, 1989).

The impact of tagging both green (*Chelonia mydas*) and loggerhead (*Caretta caretta*) turtles during their nesting activity was studied by Broderick and Godley (1999). Self-piercing plastic tags (Dalton Supplies Ltd., Henley-on-Thames, U.K.) were attached to the trailing edge of the foreflippers of females only after they had successfully completed oviposition. Behavior of tagged turtles was compared to that of untagged individuals and no effect was found on postovipositional behavior, speed of return to sea, nor on the hatching success of the resulting clutches.

Carr, et al. (1974) tagged *Chelonia mydas* for visual tracking with fiberglass-coated Styrofoam floats. These floats were "lens-shaped" (30 cm in diameter by 20 cm deep) and were attached on 24 m lines. A 3-volt flashlight bulb was attached to each float and powered by batteries imbedded in the Styrofoam. A fiberglass mast was also attached and was topped by an orange pennant. No adverse affects were found for this floatation device.

Turtle nests, such as those of *Malaclemys terrapin*, are often flagged for further monitoring and raises the question as to whether this might attract predators to the site (Burke et al., 2005). The authors found that marking Diamond-backed Terrapin nests with flags did not increase predation rates by raccoons (*Procyon lotor*) and that human scent decreased predation rates. Swingland (1978) found that Aldabran tortoises marked with discs were much more wary of humans and left the marking are immediately. This change in behavior was not thought to have a significant impact on long term studies (Swingland, 1978).

BRANDING AND PAINTING

Woodbury and Hardy (1948) used a white-hot wire to brand the carapace plates of *Gopherus agassizii*. The scutes were all assigned numbers or letters in order to obtain individual recognition codes. If scutes were burned too deeply, complete regeneration took place, and if burned too lightly the scar would wear off in a few years. Woodbury (1956) added that using a 12 to 14 gauge wire in a quick burn yielded the best results. Clark (1971) branded the plastron of one *Pseudemys scripta* into the underlying bone and found no evidence of regeneration or infection during the 36 days the animal was observed. According to Lewke and Stroud (1974), freeze branding has been successfully used on *Chelonia mydas*, but no details were given. R. K. Farrell (personal communication) reported unsuccessful use of a laser in marking a turtle.

Woodbury and Hardy (1948) also used a variety of colors of paint to mark carapace scutes of *Gopherus agassizii*. However, they found this to be less permanent than branding. Medica, et al. (1975), studying the same species, painted the last vertebral scutes and then repainted them with a different color each year in addition to using scute notching for a permanent mark. In addition to the use of radioactive tags, Bennett, et al. (1970) painted numbers of the carapace of *Kinosternon subrubrum*, *Deirochelys reticularia* and *Pseudemys scripta*. Bayless (1975) painted numbers on the carapace of *Chrysemys picta* which needed to be repainted each year because they were shed with the carapace scutes. Zoo collections of tortoises such as *Geochelone gigantea* are also often painted with numbers for easy identification (Ashton, 1978). Numbers painted on the carapace allow turtles to be identified without recapture and with little disturbance.

Burger and Montevecchi (1975) marked the plastrons of nesting *Malaclemys terrapin* with washable ink so that they would not be recounted in a census while they were on land for nesting. In a study of nesting success, the eggs were numbered with a permanent ink felt pen (Burger, 1976).

TRAILING DEVICES

Although the trailing device used by Stickel (1950) for *Terrapene ornata* was not the first, it did have tremendous improvements over earlier methods. She first made a metal housing from a six ounce can (85 x 62 mm) which fit smoothly over the caudal end of the carapace, being neither wider nor higher than the shell itself. A spindle was fashioned from an iron bolt and suspended by two wire hooks soldered to the housing. The core of a wooden spool was reduced in size so that it would then hold about 550 yards of No. 80 white thread. The spool was mounted on the spindle and the thread strung through a wire loop soldered at the rear of the housing. The entire device was attached to the carapace with waterproof adhesive tape. Spools were wound mechanically with an adapted electric mixer and could be changed relatively easily in the field. The tape needed to be replaced occasionally due to weathering. Stickel found no evidence that this device changed the behavior of the turtles.

LemKau (1970) attached a "thread trailer-radio transmitter packet", weighing about 80 g, to the carapace of *Terapene carolina*. Another modification of Stickel's (1950) technique was developed by Reagan (1974) in order to avoid any interference with mating. He used 35 mm film canisters to hold a custom lathed wooden spool on which was sound 275 m of No. 50 nylon thread. The end of the thread was passed through a hole in the canister and the unit was attached to the caudal end of the carapace with waterproof adhesive tape. The end of the string was tied to an object at the release site to begin the trail. Reagan also found this method more advantageous than the old technique in that it was tangleproof, weatherproof and only weighed 12 g as compared to 55 g.

A U.S. source of thread bobbins (Culver Textiles, P.O. Box 360, West New York, New Jersey 07093, USA, 800.526.7188) was provided by Wilson (1994) along with a simple method of encapsulating the bobbins with little bulk added. Wilson pulled about one meter of thread from the bobbin (small at 1.8 g, large at 4.5g) and then wrapped it in clear plastic wrap which was twisted at the bottom and taped to the side. Then, holding the projecting wrap with a forceps the package was dipped in Plastic Dip® (available in hardware stores for dipping tools) and placed on waxed paper to dry. After drying the flat edge next to the paper makes attachment to animals (such as *Kinosternon baurii*) relatively easy (Wilson, 1994).

An alternative to the "axle-type" trailing device was devised by Scott and Dobie (1980) in order to reduce the problem of friction by eliminating the need for an axle. A light-weight polyester thread was wound on a Styrofoam spool and mounted lengthwise in an empty 36 mm aluminum file canister to the posterior portion of the carapace of *Kinosternum subrubrum*. The device was fastened onto the turtles with three small wires and the canister enclosed a mechanism similar to the thread-release of commercial fishing reels. The spool was fastened inside the canister with round-headed machine screws and an eye ring on the inside of the canister lid created a smooth passage for the trailing thread.

Fluorescent powder (JS-CH3020, type 300 from Radiant Color, 2800 Radiant Avenue, Richmond, CA 94804, USA, 415.233.9119) was used to track adult *Gopherus polyphemus* by Blankenship et al. (1990). A fine nylon mesh pouch with 5 cc of the powder was attached to the caudal marginal scutes with cotton twine so that the pouch could drag along the substrate behind the tortoise. Each day's movement was tracked at night with a portable UV light ("Woods Light", from Henry Schein Inc., 5 Harbor Park Drive, Port Washington, NY 11050, USA, 800.872.4346) and the trail marked with flagging for later mapping over the six day study period (Blankenship et al., 1990).

The use of fluorescent pigment powder to locate movements and nesting sites in *Testudo graeca* was found to be more productive than the use of thread trailing (Keller, 1993). Four colors of Fiest Daylight Fluorescent Pigments, series E (Swada Co., London) were impregnated into a patch of rabbit fur glued on the plastron. Yellow, orange and red pigments were the best detected at night using a UV-light source. The fur patch impregnation of powder was renewed each day and most females were followed for about 7 days, covering an average of 120 m/day between egg detection via radiography and oviposition. The pigment trail documented minor digressions along the loops and back and forth movements better than a thread trail. In addition resting spots could be better distinguished. Trails can be lost after a couple days of heavy rains. No toxicity of the pigments in adults or hatchlings was found. Hatchling *Emydoidea blandingi* were dusted in a canister of fluorescent powder, avoiding the facial mucosa, but rubbing it gently on the plastron, leg pockets and front leg scale beds (Butler and Graham, 1993). When replaced at the nest site the hatchlings seemed to move normally and were easily tracked with a UV light at night. Reapplication of powder was done when necessary over several days until they reached wetland habitat.

PATTERN MAPPING

McDonald, et al. (1996) successfully used photographs of the pineal spot ("pink spot") on leatherback turtles (*Dermochelys coriacea*) for individual recognition over a period of several years. Accuracy of the technique improved slightly with experienced observers over novice researchers. In addition, rinsing the head free of sand, avoiding glare from camera flash and standardizing the camera distance and angle all improved results. The authors recommended using this technique in addition to flipper tags since it was not 100% reliable (McDonald, et al 1996). McDonald and Dutton (1996) suggest that since flipper tag retention is low for leatherbacks that the tags always be used with PIT tags and photoidentification to estimate sizes of nestling populations and studying life history profiles.

The feasibility of using plastron markings in juvenile *Glyptemys insculpta* for identifying individuals was studied by Cowin and Cebek (2006). Volunteers attempted to match plastron photos taken one year apart and achieved correct matches only 29% of the time in the first trial. The patterns of clutch mates were found to be more similar to one another than those from different clutches making sibling identification very difficult. Photographs taken six months after the initial images were matched more successfully six months later suggesting that plastron marks in the wood turtles become more stable with maturity. The authors conclude that this technique is not reliable for identifying young *Glyptemys insculpta*.

LIZARDS

TOE AND SCALE CLIPPING

Toe clipping is the most popular technique used in marking lizards. The method used by Tinkle (1967) for *Uta stansburiana* is the most commonly cited and involves clipping up to four toes, but no more that two per foot and never adjacent ones. Fingernail clippers are often used to remove the digits. The numbering system is illustrated in Figure 7A and, for example, the removal of toes 4, 8 and 20 would give the lizard a code number of 4-8-20. Medica, et al. (1971) described a different numbering system (see Figure 7B) which is similar to that used for salamanders and frogs. This technique cuts at least one toe from each foot to eliminate the problem of a lizard having lost a digit(s) accidentally being mistaken for a marked animal. Natural toe loss in some Australian skinks has been reviewed by Hudson (1996) and found to average about 20% in terrestrial species and 16% in scansorial species. This suggests that in some species natural toe loss could cause some confusion when using toe clipping as a marking technique.

Woodbury (1956) suggested lettering the feet, numbering the toes and also giving the sex for an individual an identification code. A1D2B& would indicate that the first toe of the left front foot and the second toe of the right hind foot of this male were removed. Many studies use a more visible mark, such as the paint symbols discussed below, in addition to toe clipping in order to reduce the need for frequent recapture.

Using a system which assigns letters (A, B, C, D) to the feet and numbers (one through five) to the toes on each foot, Waichman (1992) could mark up to 959 individuals when clipping up to three toes (see Fig. 1C and Table 1).

The possibility of toe clipping having a harmful affect on lizards was raised by Woodbury (1956). The impact of toe clipping on the arboreal *Anolis carolinensis*, which has pad bearing toes was assessed by Bloch and Irschick (2005). They found a 40% decrease in clinging ability when two toes were clipped and a 60% decline when four were clipped. In another study, Borges-Landaez and Shine (2003) found that toe clipping had no affect on the average or maximum running speeds of *Eulamprus quoyii* in Australia. Similarly, Dodd (1993) found that toe clipping had no immediate or permanent impact on sprint performance in *Cnemidophorus sexlineatus*. Rodda et al. (1988) found that handling and toe clipping of hatchling *Iguana iguana* led to their selection of higher sleeping perches in the vegetation. In addition they found some evidence that handling related to marking impacted the home range size of the iguanas. In as study of the effect of toe clipping on *Hemidactylus turcicus*, Paulissen and Meyer (2000) found no significant impact on survival or wall-running when no more that one toe was clipped per foot. They recommend no more toes than this be clipped in arboreal or wall-dwelling geckos until studies of additional clipping are performed (Paulissen and Meyer, 2000). Langkilde and Shine (2006) monitored plasma corticosterone levels in lizards to measure stress in *Eulamprus heatwolei* and found toe clipping had little impact on these levels while the use of microchip implantations (see PIT tags below) was more stressful (elevated blood levels for 14 days).

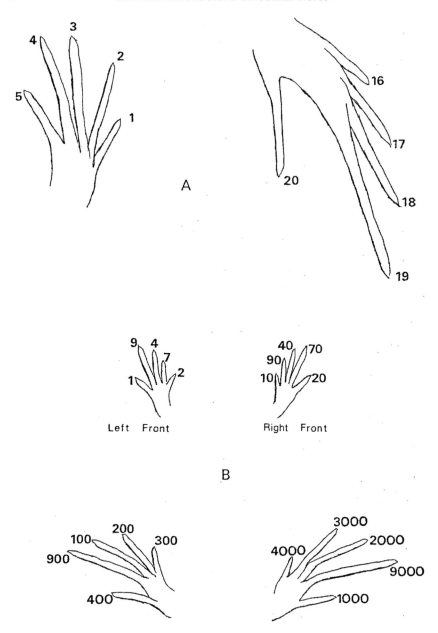

Figure 7. A. Toe numbers for the left forefoot and right hindfoot as taken from Tinkle (1967). The right hand of the lizard would bear numbers 6 through 10 from the inside out and the left foot would have toes numbered 11 through 15 from the outside in.

B. The toe numbering system for lizards as taken from Medica, et al. (1971).

Crest scale clipping and removal was used on *Iguana iguana* by Rodda et al. (1988) by removing three scales with scissors to delineate three marking sections on the crest, spaced 11 scales apart. Then, the 10 scales within each section were designated as 0 through 9 and one of these was removed to give the animal an identification number. This allows up to 999 individuals to be so marked and the scale could be repositioned or adjusted in accommodate any naturally missing scales. Subsequent scale loss did result in ambiguities which were minimized by having additional notes on tail regeneration and other characteristics. This technique was not as reliable with hatchlings as some regenerated scales and the clipping was more time consuming and prone to errors.

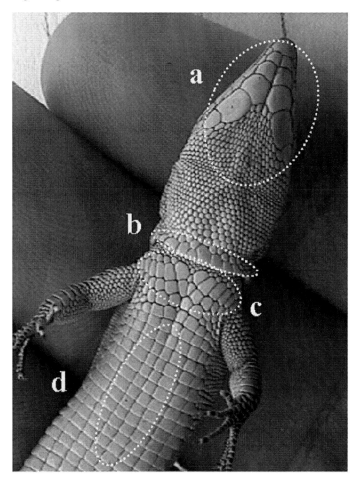

Figure 8. Photograph of venter of *Lacerta perspicillata* from Perera and Perez-Mellado (2004). Characteristics used for identification were: a) chin-scale symmetry, b) collar scales, c) scale arrangements in chest area, and d) position of longitudinal and transverse scale rows.

INTEGUMENT PATTERNS

In a field study with a small number of individuals of *Anolis aeneus*, Stamps (1973) found she could recognize individuals by their distinctive patterns coupled with tail regenerations in various stages. Carlstrom and Edelstam (1946) reported the successful use of black and white photographs to record unique individual dorsal patterns in *Lacerta vivipara* and throat patterns in *Anguis fragilis*. *Lacerta perspicillata* were successfully identified by Perera and Perez-Mellado (2004) using digital photographs of ventral scalation (see Figure 8). Of the 53 lacertids identified using this technique, 100% were verified as being correct when checked against their toe clipping marks. Perera and Perez-Mellado (2004) found that using black and white printed digital images made this technique "only slightly more time consuming and expensive than toe clipping" with the advantage of being less permanent and invasive.

In an extensive study of hatchling and adult *Iguana iguana*, Rodda et al. (1988) found that capture and handling in the field disrupted some natural behavioral patterns. Therefore, they developed a method of using various character states (length, attitude, curvature and tip type) of the dorsal crest scales to be able to successfully identify individuals at a distance without capture. A variety of other features were recorded which helped confirm visible identifications in a wide variety of field conditions. For example, over 90% of the iguanas had subtympanic scales with a distinctive pattern and could be used easily as a single trait to identify about 50% of the individuals. This technique worked well with this relatively large species in part due to its low activity level.

BRANDING AND PAINTING

Clark (1971) used heat branding successfully on *Anolis carolinensis* and *Phrynosoma cornutum* and speculates that branding is preferable to toe clipping in that locomotion is probably less affected and that branded numerals are easier to read than are toe clip formulae. Clark's technique has been described for anurans above. R. K. Farrell (personal communication) has had some excellent results in freeze branding iguanas with marks persisting through several skin sheddings.

Tinkle (1967) painted the adult *Uta stansburiana* with colored insignia (see Figure 9A) in order to identify individuals without recapture. Some confusion was caused by the shedding of skin and lizards were periodically recaptured and repainted. Young individuals were marked only with a small spot of paint on the back between the hind legs which did not allow individual recognition. Medica, et al. (1971) used Testor's model paint (blue, yellow, light green, pink and white) to mark lizards. They put the paint in nail polish bottles and used short (1/4 inch) thinned out brushes. The symbols used by Medica et al. included such things as neck bands, longitudinal stripes, transverse bands at midbody, arrows, plus signs and dots. Jenssen (1970) painted spots with quick drying paint on the backs of *Anolis nebulosus* (Figure 9B). Females were spotted with yellow paint and males

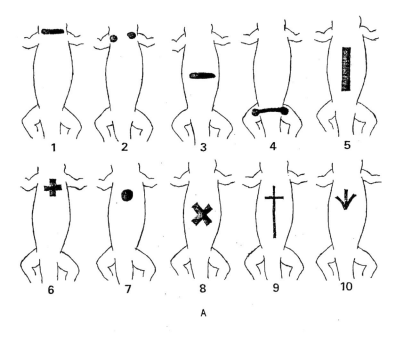

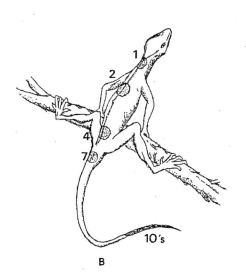

Figure 9. A. Paint symbols used by Tinkle (1967) in various combinations for individual recognition.
B. The paint spotting system developed by Jenssen (1970); see text for details.

with orange. By using combinations on the four dorsal numbers (1, 2, 4 and 7), all numbers between 1 and 9 were obtainable. The tens column was designated by painting the tails different colors (white for 10's, green for 20's etc.).

Reflective paint (Scotchlite® Brand Reflective Liquid, Series 7200, 3M Co., St. Paul, Minnesota) was used to mark *Iguana iguana* by Rodda et al. (1988). The 1 cm diameter spots remained on the lizards about two months (until ecdysis) on the average and were used primarily for night identification as they reflected spotlighting brilliantly. In the daylight the paint was dull and not as useful for identification. Depending on conditions, these reflective spots could be spotted up to 100 m in the field which resulted in high (96%) rates of resighting.

Vinegar (1975) used a variety of color combinations of model airplane paint to mark the tails of *Sceloporus undulatus*. She avoided using an excessive amount of paint in order to reduce any impact the paint might have on survivorship. In another study of *S. undulatus*, Jones and Ferguson (1980) found no evidence of increase predation on those marked with a spot of airplane paint at the base of the tail. Simon and Bissinger (1983) designed a study of the impact of paint marking on survivorship in *S. jarrovi* and found that the color used and painting in general "did not affect survivorship". Since these species of *Sceloporus* are sit-and-wait predators, the results may be different for species that are more active foragers such as in *Cnemidophorus* so that might be considered when using paint marks and deserves further study (Simon and Bissinger, 1983).

Some more recent studies have used xylene-based paint pens to mark lizards. Boone and Larue (1999) did a pilot study of the impact of this technique on the survival of *Uta stansburiana*. Of 21 lizards painted in a lab study, five (four juveniles and one adult) over a period of two weeks while none of the 14 lizards in the control group died, a statistically significant difference. In addition, sleeping behavior was also impacted in that fewer of the marked lizards burrowed underground at night as is normal which may, in the field, expose them to more predation or thermal stress. Boone and Larue (1999) recommend not using xylene based paints for marking lizards or other animals.

Stebbins and Cohen (1973) used a purple indelible pencil to color the base of the hind leg in *Sceloporus occidentalis* and thus indicate whether individuals had been parietalectomized (left leg) or sham-operated (right leg). Similarly, Henderson (1974) used a felt-tip pen to write black numbers on the sides of *Iguana iguana* and reported that these marks lasted several weeks.

Fluorescent powder (available in six colors) was used to track *Xantusia riversiana* at night by dipping them in a plastic bag containing about 50 ml of powder (Fellers and Drost, 1989). Trails could be followed reliably for about two nights. Powder was purchased from Radiant Color (2800 Radiant Ave., Richmond, California 94804 USA; series R-103-G) and was detected using an ultraviolet light (Blak-Ray, longwave, ML-49) from Scientific Marking Materials (P.O. Box 24122, Seattle, Washington 98124 USA). Use of this technique with diurnal lizards was discouraged as the bright colors might increase risk of predation or impact behavioral interactions.

It is unknown how colored marks affect the behavior of congeners toward marked specimens in species for which body colors have seasonal social significance.

TAGGING

Minnich and Shoemaker (1970) marked *Dipsosaurus dorsalis* with colored Mystik cloth tape in addition to toe clipping. Different arrangements of the various colors of tape were placed around the base of the tail. This technique has also been used to mark *Uma scoparia* (Minnich and Shoemaker, 1970). In a similar technique, Zwickel and Allison (1983) marked *Emoia physicae* with pressure sensitive rip-stop nylon tape (Coghlan's Ltd., Winnipeg, Canada). After wiping the dorsum with 95% alcohol, a 5 X 10 mm piece of tape was attached and then color coded with acrylic paint which was allowed to dry before release. Lizards were toe clipped for permanent identification and were remarked when tags were lost which was only common when skin was being shed (Zwickel and Allison, 1983).

Uma notata have been tagged with a small piece of foil attached to a 30 cm, light string which was tied around the lower abdomen (Deavers, 1972). This tag allowed the measurements of burial depth of the lizards at night. Judd (1975) used a similar tag to locate buried *Holbrookia propinqua* for body temperature readings. He used a 5 cm square piece of aluminum foil and a 1 m length of red thread.

Nocturnal activity of anoline lizards was studied by Clark and Gillingham (1984) by attaching glowing tubes to their dorsum with Duco® cement. Micropipette sections (1.8 X 30 mm) were filled with a phospho-luminescent liquid from a Cyalume® lightstick (American Cyanamid Co., Bound Brook, New Jersey) with a hypodermic syringe and then sealed with Seal-Ease® tube sealant (Clay Adams, Parsippany, New Jersey, USA). After allowing the glue to dry 5 minutes the anoles were released and could be observed from up to 30 m for up to 6 hours after dusk. In 20 trials all lost the tag within 24 hours with no apparent damage to the lizard.

Rao and Rajabai (1972) tagged agamid lizards (*Sitana ponticeriana* and *Calotes nemoricola*) with different shaped, colored aluminum rings. These rings were placed around the thigh and caused no apparent hindrance. Henderson (1974) tagged Iguana iguana by tying small "jingle bells" around their necks with fishing line!

Using beads approximately 2 mm in diameter, strung on nylon monofilament line, Rodda et al. (1988) tagged *Iguana iguana* on their mid-dorsal flap of integument. This crest was pierced with a hypodermic syringe needle so the nylon line could be threaded through and then secured by melting the ends of the line after the beads were inserted. Enough slack in the line was left to allow for growth and the tag was tested for strength by pulling on the beads to be sure they would not slip off the line. Some small hatchlings may have lost these tags, but juvenile skin was much tougher and no problems were noted.

In a study of *Uma inornata*, Fisher and Muth (1989) modified a bead tagging technique used by Nace and Manders (1982) for amphibians. The stages in this marking process are illustrated in Figure 10 (Fisher and Muth,

1989). Plastic jewelry beads (2mm X 2.5 mm) were strung along 15 cm surgical steel monofilament strands in the lab (Fig. 10A). A 22 gauge hypodermic needle is inserted through the ventral side of the tail "at a point distal to the male's hemipenis and lateral to avoid the caudal vertebrae" and the strand in then threaded through the needle (Fig. 10B). The needle is removed and then reinserted through the dorsum of the tail and the loop at the end of the strand just distal to the first penetration. The strand is then pulled back through the needle dorsally (Fig. 10C). After removing the needle, the steel strand is "bent back through the loop, pulled tight, then cut close to the loop leaving the beads securely fastened dorsolaterally" (Fig10D). The authors recommend practicing on preserved specimens of the species before using it in the field and that once practiced a person can apply the tag in about five minutes. Only lizards larger than 6 cm were tagged with the beads and no adverse effects were reported. Fisher and Muth (1989) reported that this technique has also been used successfully with *Phrynosoma mcalli*, *Gambelia wislizenii* and *Dipsosaurus dorsalis*.

Since teiid lizards are difficult to successfully mark with paint due to their small smooth dorsal scales and burrowing habits, Paulissen (1986) developed a technique of gluing plastic bird bands to the tails of *Cnemidophorus sexlineatus*. The bands (National Band and Tag Co., Newport, Kentucky USA) were cut so only a single complete ring was used with no gap present between the ends, having an internal diameter of about 3/16 inch. After sliding the band up the tail until it was tight and sized to be about two-thirds of the distance up the tail, it was secured with two drops of Duro® superglue. Two to three bands of different color could be used in order to provide a variety of color combinations. With the use of multiple bands it is recommended that they be spaced so that the tail may move freely. Bands are often lost with ecdysis so replacement may be necessary. If the bands begin binding the tail severely it is recommended that they be replace with larger ones so as to minimize tail deformity. To minimize this problem with juvenile lizards undergoing rapid growth, no more that two bands were used on specimens with a SVL of less than 40 mm. Average band life was reported to be 26.4 days (N = 23, range of 4 to 63 days) by Paulissen (1986).

Johnson (2005) used bee marking kits to tag four species of lizards (*Anolis cristatellus, A. gundlachi, A. krugi,* and *Sceloporus undulatus*) for short term studies, i.e. those that can be done between successive skin sheddings. The kits were obtained from The Bee Works of Orillia, Canada (www.beeworks.com), each containing numbered cardboard dots in five colors, phial glue and an applicator. The dots were found to be useful in about 85% of the lizards under study over a three-week period. The technique is relatively inexpensive, non-invasive and provides a highly visible mark.

Visible implant elastomer or VIE (Northwest Marine Technology, Shaw Island, Washington, USA) was injected under the ventral skin around the leg joints of *Hemidactylus turcicus* by Daniel et al. (2006). This mark was successfully retained during ecdysis, while the incidence of loss with surface marking techniques (superglue dots and alphanumeric tags) was virtually 100% with shedding. The authors conclude that even though the VIE requires more cost and effort than surface marks, these factors are out-weighed by the long retention of the injected elastomers.

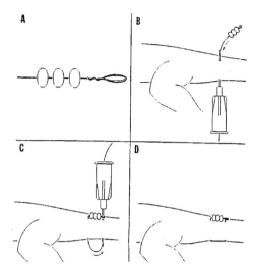

Figure 10. Steps in the lizard marking technique from Fisher and Muth (1989). See text for details.

SNAKES

SCALE CLIPPING

Scale clipping with scissors or toenail clippers is by far the most commonly used method of marking snakes. Many investigators still follow the technique developed by Blanchard and Finster in 1933. They simply cut pieces from the subcaudals leaving "permanent" scars. However, regeneration would be almost complete if the piece removed was nearly the entire scale and integument down to the muscle layer with sharp-pointed scissors. They reported no adverse effects on the snakes.

Blanchard and Finster (1933) numbered the subcaudals on each side beginning at the proximal end of the tail. Therefore, if the fourth and sixth left and forth right subcaudals are clipped, the individual was coded 4, 6-4. They usually scarred three scales per snake and only up to scale number 20. Since some snakes have a small postanal scale which could cause confusion as to which is numbered one, the most proximal subcaudal clipped on either side was always an even number.

Hirth (1966) marked hatchlings and some juveniles of *Masticophis taeniatus* and *Coluber constrictor* and all ages of *Crotalus viridis* with ventral scale clips as part of a hibernation study. While the details of the technique were not explained it can be assumed to be similar to that describe above and was apparently effective.

Carlstrom and Edelstam (1946), working with *Natrix natrix*, criticized the technique of Blanchard and Finster (1933) in that regeneration was a problem and that it was difficult to mark hatchlings. Woodbury (1956) cites regeneration of clipped scales in 4 or 5 years as reported by Conant for *Elaphe obsoleta*. Weary (1969) found three disadvantages to the use of Blanchard and Finster's technique in marking *Storeria occipitomaculata* and *Thamnophis sirtalis*: 1) Several minutes were required to remove carefully a complete caudal scale; 2) Despite precaution blood was frequently drawn, allowing possible infection; and 3) Clipping proved to be very difficult on young snakes less than 10 cm long.

Brown and Parker (1976b) have described a ventral scale clipping system, which they feel fulfills more of the criteria for an ideal mark (see Preface) than does Blanchard and Finster's (1933) method. They see it as an important improvement because ventrals are "larger and easier to clip than subcaudals" and scars in that area cannot be lost by tail breakage. Brown and Parker have also devised a serial enumeration system (see Figure 11) which yields a whole number for each specimen and is "far less confusing" that the number code of Blanchard and Finster.

Brown and Parker (1976b) used sharp-pointed scissors of appropriate size and quality and described the procedure as follows:

(1) Insert a top of the scissors under the posterior edge of the scute to be clipped, push it forward beneath the width of the entire scute, and cut. Make two such incisions, one on each side of the intended block to be excised.

(2) Insert the scissors under and across the top (anterior edge) of the scute and make a third cut transversely to remove the entire section. Depth of the cuts is important: the entire excision should include all layers of the skin (including the dermis) to expose the underlying ventral musculature. Muscles should be visible and the section removed should be large enough to cover half of a given ventral. In cases of dual numerals involving the same scute (55, 66, etc.), clip all the way across and leave some lateral traces of the scute so that its position can be counted after healing. There is an apparent tendency for adjacent scutes to "invade" an excised area (see Fig. 11) or to adpress against an excision zone after removal of an entire scute. Numbers involving adjacent scutes on the same side (e.g., 190's, 1900's) should be omitted.

The authors did not observe direct adverse effects of this technique on snakes. However, they did relate scale clipping and/or handling as causes for increased over-winter mortality and weight loss in newly-marked *Masticophis taeniatus*. Their studies found no similar affects with *Coluber constrictor* or *Pituophis melanoleucus*. Brown and Parker reported these marks permanent for at least four years and often (92% of the time) shed skin from clipped *Coluber* would be precisely identified.

BRANDING AND PAINTING

Woodbury (1956) reviewed his tattooing technique for snakes which used a portable, battery powered tattoo outfit. Numbers were made on any light areas where pigmentation would not obscure any portion of the mark. In order to minimize misreading numbers upon recapture, the free ends of the numerals 2, 3, 4, 5, 6, and 9 were "made extra long" so that they were better distinguished from each other and from the numeral 8, having no free ends. Woodbury found this mark permanent and with no adverse effects after the recovery of the snake from the tattooing operation.

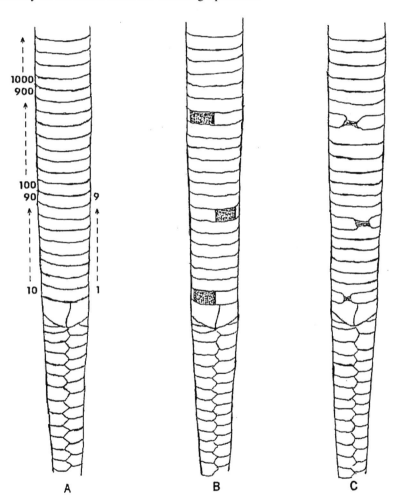

Figure 11. The ventral scale clipping system of Brown and Parker (1976b).
    A. The enumeration system.
    B. Freshly clipped snake No. 718.
    C. Scars on the same individual (No. 718) three years after marking.

Storeria occipitomaculata and *Thamnophis sirtalis* have been branded with a 50-watt pyrographic needle which can be purchased in hobby shops (Weary, 1969). The needle was filed to give a sharp cutting edge and applied briefly to burn through scales without causing bleeding. No regeneration was observed over a two-year period. If no house current is available, Weary suggested using the more expensive American Beauty soldering pencil, model B2000-L, powered by a 12 volt rechargeable battery.

Clark (1971) branded several species of snakes using Hoskings Chromel "A" resistance wire (The Malin Co., Cleveland, Ohio USA) which is 20% chromium and 80% nickel. A 20 mm piece of wire B and S gauge No. 24) fixed in a handle was used and heated with either a Bunsen burner or a small, portable propone torch. Brands were placed on the ventrals and/or subcaudal scutes using a numbering system similar to that of Blanchard and Finster (1933) which is described above. The obliteration of the brands by regeneration was reduced by including the first row of adjacent dorsal scales in the mark. This resulted in a healed area being covered by either a single scale, "as though the ventral (or caudal) and adjacent dorsal had fused" or the normal division between the two scales was out of place. In addition, the regenerated portion may have a change in pigmentation.

The wound produced by Clark's (1971) brand forms a scab within a few days and it not an open sore. Sometimes, scales may be sloughed off before complete recovery by the underlying integument, leaving an open sore. Even in these cases, Clark believed survivorship to be less influenced by branding than by scale clipping. He also commented that the increased survivorship of branding over clipping may offset the inconvenience of having added equipment in the field. Clark had data suggesting that healing of the brands occurred more rapidly and with fewer complications in the field than when snakes were confined in the laboratory.

Lewke and Stroud (1974) marked two *Crotalus vividis* and seven *Pituophis melanoleucus* by freeze branding. A super-chilled, copper branding instrument was used to alter the normal pigmentation by destroying the chromatophores, but not permanently changing any other part of the integument. Bar and angle shaped irons (0.5 cm by 1.5 cm by 1.0 cm deep) were fashioned from a copper bar or tubing and soldered to sturdy wire angles which were 15 cm long. The authors experimented with three coolants:

(1) An equal part crushed dry ice – 95% ethyl alcohol (-70° C).
(2) Liquid Freon 12 (boiling point, -21° C).
(3) Liquid Freon 22 (boiling point, -41° C). Both Freons obtained from Virginia Chemicals, Inc. Portsmouth,VA in 500 g pressurized cans.

The coolants were placed in insulated containers and the irons were cooled until boiling stopped at the iron surface, "indicating that the metal had reached the temperature of the coolant". The brand site was swabbed with 95% ethyl alcohol to increase conduction and the branding iron was held against the integument for 5 to 30 seconds. Only the dry ice-alcohol coolant produced successful marks which were best at 20 to 30 second exposures. These marks have been observed for about two years and produced no adverse effects in the snakes. The brands may induce molting as all the snakes used in this study molted within three weeks of marking.

Both types of Freon were used successfully by Lewke and Stroud (1974) when applied for 5 to 20 seconds with L- or bar-shaped synthetic foam sponges fastened to wooden dowls with a silicon adhesive. Applying the Freon 22 by direct spraying over a styrofoam stencil for 1 to 2 seconds produced visible brands, but the shapes were usually distorted because of leakage around the stencil. The freon coolants were used primarily because they are more convenient to carry in the field.

The primary disadvantage of freeze branding, according to Lewke and Stroud, is that the mark is not visible until the first molt after branding. Additionally, the color background of the snake must be considered and a minimum branding iron size would be reached "because the heat sink available for cooling is a function of the mass."

R. K. Farrell (personal communication) has used a ruby laser to permanently mark king snakes and rattlesnakes. This work was done as a peripheral study to projects which concentrated on the markings of crabs and fish. This technique was developed for the rapid marking of extremely large numbers of individuals.

Disposable medical cautery units ("Aaron Medicla Change-A-Tip cautery units;" Aaron Medical, St. Petersburg, Florida 33710, USA; www.aaronmed.com) were used to mark snakes by Winne et al. (2006). Two temperature classes of these units are available and both were used successfully. High temperature units (1204° C costing U.S. $25) use either 2 C or 2 AA alkaline batteries depending on handle size and the low temperature option (704° C at U.S. S20) requires one AA battery. These units are long lasting, but do require regular battery changes and fresh batteries deliver higher temperatures so caution should be used. An example of a heat-branded snake is shown in Figure 12 with most marks lasting at least two years on all sizes of individuals. Over 200 heat-branded snakes from 15 species were recaptured over periods up to 1058 days.

Pough (1966) marked three species of *Crotalus* "by painting an identifying number on the basal rattle segment with quick-drying waterproof paint (Testor's Butyrate Dope)." A similar technique was reported by Brown et al. (1984) for rattlesnakes by dabbing Liquitex® acrylic paint (Binney & Smith, Inc., Easton, Pennsylvania 18042 USA) on the rattle. In this case individual recognition was secured by the permanent mark of ventral scale clipping (Brown et al., 1984) and paint marks remain visible on *C. horridus* for up to four years. *Masticophis taeniatus* have been painted on the head and neck with a color code for individual recognition (Parker, 1976; Bennion and Parker, 1976).

TAGGING

Hirth (1966) tagged adult *Masticophis taeniatus* and *Coluber constrictor* with numbered Monel® tags which were "clamped into the corner of the mouth." No details concerning this technique were given and it apparently has not been widely used by other investigators.

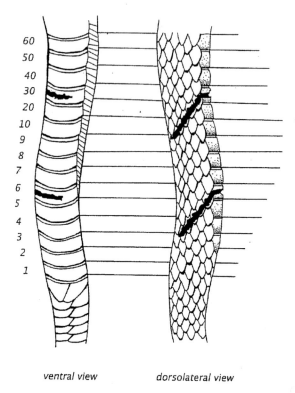

Figure 12. Illustration of heat-branded snake with ID # 36 from Winne et al. (2006). Each brand includes the anterior portion of the ventral scale and extends dorsally onto adjoining lateral scales. Drawn by Rebecca Taylor.

The Buttoneer® (as discussed in the section on turtles) has been used to mark snakes by inserting a plug into the caudal musculature through the lateral region of a subcaudal scute (Pough, 1970). Plugs should be positioned so that the bulb rests tightly against the snake's integument. Difficulty may be encountered in marking smaller snakes (25 cm) because of their thin tails, but using a Buttoneer® with a shortened tip may alleviate this somewhat.

Pendlebury (1972) tagged *Crotalus viridis* with a pair of colored vinyl discs which were fastened "through the dorsal lobe of the second proximal segment of the rattle" (see Figure 13). The discs were 6.0 mm in diameter and 0.5 mm thick and were punched from large sheets (available at most plastic supply houses) with a paper punch. A 50 mm length of monofilament nylon fishing line was passed through a hole made with a needle in the center of the pair of discs. One end of the line was melted into a bead and the other end knotted to secure the discs. Upon capture of the rattlesnakes, a No. 21 hypodermic needle was passed through the rattle as illustrated in Fig. 12A. The knotted end of the line was then cut and one disc removed. This end of the line was then threaded through the needle, the needle was removed and the disc was replaced over the loose end of line (Figure 13B and C). The cut end of the line is then beaded with a match and the snake may be released (Fig. 13D). Pendlebury reported that these tags lasted at least 15 weeks and believed them to be essentially permanent in that the more distal segments of the rattle are more subject to wear. Young rattlesnakes might be tagged on the "button" which would run more risk of loss. The tag was found to be readily observable from distances of 6.5 to 30 m (with binoculars). Using 10 colors of plastic discs, 100 individuals could be distinctively tagged. The number of molts which have occurred since tagging can be determined by (N-1), where N is the number of segments craniad to the tagged one. Pendlebury reported no adverse effects of this technique.

The use of fish tags on *Crotalus adamanteus* was evaluated by Smith (1994). These plastic tags were from Floy Tag and Manufacturing (P.O. Box

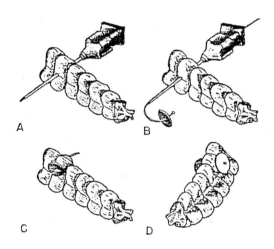

Figure 13. Illustrations of the technique from Pendlebury (1972) for tagging rattlesnakes. See text for details.

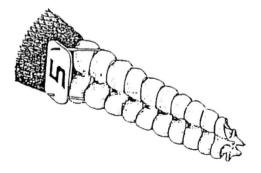

Figure 14. Illustration from Smith (1994) of a plastic fish tag fastened on the proximal rattle of *Crotalus adamanteus*.

85357, Seattle, Washington 98105 USA) measuring 1.5 X 0.5 cm. The tag was tied between the two proximal rattle segments using monofilament line (see Figure 14) with the ends being knotted and secured with Superglue™. Tags were marked with sequential numbers on one side and the address and telephone number of the investigator on the other side. Smith (1994) found that these tags were superior to those used by Pendlebury (1972) in that the contact information could be included on the tag, but the life-span of the tag was not determined.

Stark (1984) tagged rattlesnakes using fine stainless steel wire, colored sequins and epoxy. Pink (for females) or blue (for males) sequins were threaded onto short pieces of the wire with each sequin being stenciled with an identification number. The wire was twisted on each specimen, after they were restrained, between the body and the proximal rattle segment being sure the sequin was flush against the widest side of the rattle. An epoxy coating was then placed over the wire and sequin. These tags allowed immediate sex identification of individuals in the field from 15 m with the naked eye.

A color bead tag for snakes of all sizes was developed by Hudnall (1982). A glass bead (variety of colors available in craft stores) is threaded on monofilament suture-line which is knotted at one end. The line is threaded onto a surgical needle and "inserted into the tail dorso-laterally and run along the backbone for at least 13 mm." After pulling the bead snug against the tail the line is cut long enough to allow the threading of an additional bead or two and the tying of a second knot. While these tags often were lost in shedding for *Sistrurus miliarius*, Hudnall suggests they will last even longer in larger snakes. No adverse impact of this technique was reported.

INTEGUMENT PATTERNS

Carlstrom and Edelstam (1946) photographed the ventral black and white pattern of *Natrix natrix*. They found this pattern to be of infinite variation, but consistent during the life history of each individual. Therefore, they described the system of individual recognition to be "as sure as the fingerprints of the police." Others have identified snakes by using damaged tails and other natural marks, usually along with traditional permanent marks such as scale clipping (Spellerberg and Prestt, 1978). For example with *Vipera berus* juveniles were photographed to record their mid-dorsal patterns by enclosing them in the groove of an entomological setting board covered with a microscope slide (Spellerberg and Prestt, 1978).

Photographs of the tail pattern variation in *Crotalus atrox* (see Figure 15) proved to be a successful method for individual identification in a study by Moon et al. (2004). For verification rattle painting, scale clipping and PIT tags were used for permanent marks. Only one of 261 tail pattern photo-identifications was in error due to similarity in appearance.

The Ellen Trout Zoo developed a technique for identifying individual snakes by saving their exuvia after ecdysis and noting any scale pattern irregularities, scars or other characteristics (Henley, 1981). After drying the

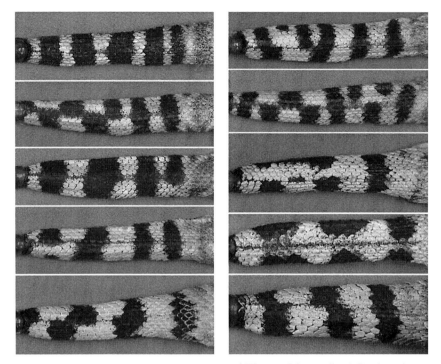

Figure 15. Variation in tail banding patterns of Western Diamondback Rattlesnakes (*Crotalus atrox*) as used for identification by Moon et al. (2004). Photographs are not to same scale.

exuvia, the identifiable portion was cut out, placed on the specimen's data card and covered with clear, adhesive plastic for future use.

Another use of scale anomalies has been developed for subcaudal variation in *Pseudechis porphyriachus* and *Liasis fuscus* by Shine et al. (1988). The number and position of entire vs. divided subcaudals recorded from the anal plate posteriorly for individual recognition. The detail of the records taken depended on the variation found and for more clarity simple counts were often supplemented with information on aberrant scales, partly divided or "creased" subcaudals and anomalies between the scale counts on the left and right sides of the tail. A simple formula was used to record the subcaudal pattern where, for example, "1E, 1D, 1PD, 2E, "D" is translated as an entire first subcaudal, divided second, partly divided third, entire fourth and fifth and the remainder divided. Notes on anomalies can be added verbally or with drawings on the data sheets. Shine et al. (1988) found unique patterns on from 32% to 77% of individuals in various populations so this technique was used with ventral scale clips. The authors recommend this technique as a backup to only scale clips, for situations where fewer clips may be preferable and neonatal snakes and small species where clipping is not practical.

CROCODILIANS

Chabreck (1965) reported on several methods which he used in marking *Alligator mississipiensis*. Toe clipping was successful and permanent, but was somewhat limited for use in individual recognition. So, toe clipping was combined with the notching of tail scutes (the rigid scales located dorsally and caudally on the tail). This method provided combinations for over 3000 distinct marks by assigning each scute a letter and each toe a number. In addition, a Monel® tag was attached to a dorsal tail scute in order to have a mark which was obvious to people not familiar with the study. This tag was of the self-piercing type, size no 681 from the National Tag and Band Co. (Newport, Kentucky 41071 USA), and was imprinted with a number and return address.

Additional techniques were tested by Chabreck (1965) and found to be less satisfactory for long term use: streamer tags, tattooing, and vinyl plastic collars. Joanen and McNease (1973) and Kay (2004) have used telemetry to tag alligators.

Stephen J. Gorzula (personal communication) has successfully marked *Caiman crocodylus* by clipping triangular pieces out of the ventral tail scutes according to a coding system. He has found the marks identifiable for two years, but has generally remarked individuals after one year as regeneration of scales does make them more difficult to read after that time. Similarly, Whitaker (1978) clipped dorsal tail scutes to mark *Crocodylus palustris* and *Gavialis gangeticus* using sterile surgical scissors in smaller animals and clean garden clippers in larger individuals. The authors suggests the application of an antibiotic to the wound which heals quickly. Hatchings and juveniles sometimes had to be clipped annually as the scutes often partially regenerated (Whitaker, 1978). Tailtips were not used as they are frequently lost, but with 12 double and 5 single scutes available Whitaker could permanently mark over 100 individuals.

Jennings et al. (1991) evaluated four marking techniques (Monel® tags, scute clips, web hole punching and toe clipping) in hatchling *Alligator mississippiensis* and found no impact on growth or survivorship. The authors suggest that overall, the scute clipping technique seems to be the best with respect to overall ease of application and ability to identify many individuals. However, other factors such as length of the study and mark retention times should also be considered in selecting a marking technique for crocodilians (Jennings et al., 1991).

Individual recognition of young *Ghavialis gangeticus* was found to be possible by using the unique pattern of tail bands and how the pattern changes on the various tail scutes (Singh and Bustard, 1976). The number of scutes on the single (distal) crested region of the laterally compressed tail ranged from 21 to 24 (N=52) and these were given numbers. The location of the black stripes relative to these scutes was noted and only eight scutes were needed to distinguish all the specimens used in this study. Singh and Bustard (1976) also report successful use of this non-invasive technique with a small study group of *Crocodylus palustris*.

PASSIVE INTEGRATED TRANSPONDER (PIT) TAGS

A new marking technique since the last edition of this circular (Ferner, 1979) is the use of passive integrated transponders or PIT tags beginning in the late 1980's. An extensive review and evaluation of this microchip marking system for amphibians and reptiles was given by Camper and Dixon (1988). The 10 X 2.1 mm PIT tags are encased in glass and encoded with an alpha-numeric code which is read by a portable reader with a hand wand (Camper and Dixon, 1988). The authors tested the tags on 95 individuals from a wide range of species and implanted the tags with a metal syringe having a No. 12 gauge needle. This baseline study experienced only one failed PIT due to a cracked glass cover, only a 6% error rate in the readings, a reading success on the first pass of the wand of 92% and a migration rate of a PIT within the animal of only 36% (with possible reduction of first pass reading rate success) (Camper and Dixon, 1988).

FROGS AND TOADS

All anurans in the Camper and Dixon (1988) project had the PIT tags implanted intra-abdominally using the implanter syringe with the needle dipped in 70% EtOH before implantation. Wounds were cleaned after injection with 70% EtOH and sealed with "Krazy Glue".

Brown (1997) inserted PIT tags into *Bufo bufo* and *Rana temporaria* by pinching a flap of the dorsum and injecting it with a sterile needle. After insertion the tag was gently massaged caudally until it was position between the back legs at the base of the spine. No tags were lost and there was no evidence of an impact on survival (Brown, 1997).

In a similar study, Ireland et al. (2003) marked *Rana catesbeiana*, *R. clamitans* and *R. pipiens* subcutaneously with PIT tags without the use of an anesthetic. Here the frogs were injected on the ventral side, caudal to the center after swabbing the site with 70% EtOH to sterilize and dry it. The ventor was pinched up about 4 mm with a forefinger and thumb for the injection and the wound was swabbed again with ethanol, no sutures or glue was needed. In all but 5% of the frogs the tag migrated on its own to the caudal lymph space between the hind legs. A field scanner (AVID Marketing, Inc., Norco, California, USA) was used for identification and the technique is believed by the authors to be the best available for marking medium-to-large sized anurans. The primary drawback to the technique is the cost of the AVID microchips (ca. US $8.00) and scanner (ca. US $1000) (Ireland et al., 2003).

After successfully PIT tagging 790 *Rana pretiosa*, McAllister et al. (2004) recommend them for use over most other currently available techniques. They inserted the 12 mm tags (Destron Fearing Corporation, South St. Paul, MN, USA; 125 kHz model) on frogs $\geq$ 42 mm by making a 2 mm incision 1 to 2 cm caudal to the eyes on the dorsum with a surgical scissors. The tags were then massaged back along the body to the base of the spine to minimize the chance of their being lost through the incision prior to healing (about 2 weeks).

Pyke (2005) provides a good review on the use of PIT tags and reports on the marking of 3000 individuals of nine species including *Litoria aurea* and *Limnodynastes peronii* in Australia. The tags used in this study were supplied by Trovan and are "individually-packaged needles inside hermetically sealed packages". Tags were inserted just behind the front limb along the side of the body without the use of anesthesia in frogs with a minimum SV length of 40 mm. The resulting small wound was sealed with Vetbond® (n-butyl cyanoacrylate adhesive) and <1% of the animals exhibited any sort of distress call during the entire procedure. As with previous studies, Pyke (2005) found little immediate impact or affect on reproduction and long-term survival with the use of PIT tags in anurans.

SALAMANDERS

A study of the impact of toe clipping and PIT tags on the growth and survival of *Ambystoma opacum* was reported by Ott and Scott (1999). Using a surgical technique, salamanders were anesthetized for about ten minutes with 2-phenoxy ethanol (30 drops/500 ml dH_2O). The tag was inserted into the body cavity through a 3 mm incision in the skin about 5 mm cranial to the hind limb on the right side of the body and sealed with New Skin® liquid bandage. Recovery from the anesthetic was done in containers of dH_2O with the head about water over a 3 to 4 hour period until the animals appeared active when prodded gently. While no adverse impacts of PIT tags were reported, Ott and Scott (1999) did point out the disadvantage of cost and marking time needed per animal over the use of toe clipping (with its own set of disadvantage as discussed earlier).

A long-term study of *Taricha torosa* in California marked 36 adults with PIT tags (AVID) which were anesthetized with MS222 (Watters and Kats, 2006). The tags were inserted into the abdomen with a syringe and the incision was sealed with New Skin® (Medtech). Individuals were recaptured up to 11 years after marking with 39% of the original sample being relocated during the study.

TURTLES

For PIT tag placement in turtles with hard shells, Camper and Dixon (1988) used an electric drill to put a 3/32" hole in the shell, 15 to 30 mm deep, into which the tags were placed with a sterile forceps after being dipped in 70% EtOH. A variety of placements in the carapace and plastron were attempted with the tags being secured in their holes using dental acrylic plugs. Some tags inadvertently entered the abdominal cavity and some others were intentionally placed there, all with no ill effects. Soft-shelled turtles had the tags inserted into the body cavity from the left edge of the plastron. All 32 of the 10 species marked by Camper and Dixon (1988) were marked successfully with little movement or loss of the tag.

Fontaine et al. (1987) recommended that PIT tags be placed intramuscularly in the left front flipper caudal to the radius of *Lepidochelys kempi* rather than in the carapace to prevent tag loss. The authors believe these tags have the potential for providing a life-time mark due to the long life span of the transponders.

In a study of leatherback turtles (*Dermochelys coriacea*), McDonald and Dutton (1996) injected PIT tags 4 to 5 cm below the skin surface directly into the shoulder muscle. New, sterile 12 Gauge 3.81 cm needles were used and the injection site was treated with Betadine® before and after insertion. Both the tags and scanners used in this study were made by AVID (American Veterinary Identification Devices), Inc. The authors found PIT tags to be far superior to external flipper tags and had 100% retention over a two to three year period in the study. McDonald and Dutton (1996) also noted that since there are several brand of tags and scanner which operate at different frequencies caution should be used in ordering equipment for long term studies. They found that AVID scanner also read both 125 KHz and 400 KHz Frearing Destron tags.

While scute notching is successful in marking many turtles, Buhlmann and Tuberville (1998) suggest that PIT tags may be more appropriate for use with small freshwater turtles such as juvenile *Trachemys scripta*. Tags were injected into one of three different inguinal areas of the body cavity in individuals ranging from 86 to 131 mm in plastron length: cranial and parallel to the shell bridge, cranial and perpendicular to the spine, or caudal and parallel to the carapace edge. The injection site was swabbed with 70% isopropyl alcohol before injecting the tag through the skin and all muscle with a 12-gauge needled covered with antibiotic ointment. The puncture wound was sealed with New Skin® Liquid Bandage (Medtech Laboratories, Inc., Jackson, WY, USA). The results of this study indicated no adverse effects from the tags and little internal migration was found by comparing X-rays taken initially and with each recapture (Buhlmann and Tuberville, 1998).

A study of PIT tag retention and migration using x-rays in *Trachemys scripta*, *Pseudemys floridana* and *P. nelsoni* was reported by Runyan and Meylan (2005). After placing turtles on their carapace the tags were injected into connective tissue and muscles between the plastron and pelvis lateral to the midline using a 13-guage needle. The insertion point was into the tissue just dorsal to the xiphiplastron, ventral to the ischium and lateral to the cloaca. The cost of the PIT tags (US $6.00 each) and AVID Inc. scanner (ca. US $400) is significantly less than that reported by Ireland, et al. (2003) mentioned above which is a positive sign for future investigators. Runyan and Meylan (2005) x-rayed eleven *T. scripta* over a span of 50 months where one tag was lost, presumable back through the injection site, due to not being placed at adequate depth. Little migration of tags was detected. PIT tags are recommended over drilling or notching the carapace because they are easier to read, resulting in fewer errors in identification (Runyan and Meylan, 2005).

Hatchling *Chrysemys picta* were marked with PIT tags inserted into the peritoneal cavity with a 1 mm internal diameter needle (Rowe and Kelly, 2005). The hatchling had a mean initial mass of 4.75 g and carapace length of 24.9 mm. The sterile small tag (1 X 12 mm, model TX1400L, Biomark Inc., Boise, Idaho, USA) was placed into the a sterilized needle which was "placed beveled-edge upwards against the taut abdominal skin midway between the [left hind] limb and the plastron." Just the bevel was then gently inserted through the skin in order to minimize the wound and contact with abdominal organs. The tag was injected into the body and the needle removed leaving just about 1 mm or the tag exposed. The tag was then pushed into the cavity with a sterile blunt probe and the incision cleaned with 10% povidone-iodine solution. After drying the incision was sealed with an antiseptic liquid (1% 8-hydroxyquinoline) bandage (New Skin®, Medtech, Jackson, Wyoming, USA). No impact on growth or survival of the 8 hatchlings was noted in this study (Rowe and Kelly, 2005).

LIZARDS

Thirty-five lizards from four species were tested with PIT tags by Camper and Dixon (1988) which were injected subcutaneously in the neck (in *Crotophytus collaris* only) or in the body cavity. Subsequent movement of the tags was reported in twelve individuals and the technique worked well overall.

In a four-year study of *Gambelia sila*, 581 individuals were marked with PIT tags (Germano and Williams, 1993). After trying a variety of injection sites, intra-abdominal injection was selected using a modified 3-cc syringe with a 12-gauge needle. Tag loss was reported at 8.4% for first recaptures and 4.1% for all recaptures (N = 558) and only three tags malfunctioned over a 3.5 year period. Twenty tags were lost during the study, 17 from those injected subcutaneously and three from those injected abdominally.

In a case study of *Eulamprus heatwolei*, Langkilde and Shine (2006) found increased plasma corticosterone levels for 14 days in lizards with microchip implantation, suggesting some level of stress to the animals. Levels were much higher in females than males, but the study did not attempt to measure any long term impact of microchip implantation on lizard fitness.

SNAKES

Camper and Dixon (1988) injected pit tags subcutaneously on the left side at the second dark band in *Agkistrodon contortrix*, ventrally beneath the subcaudals in *Mastocophis taeniatus, Pituophis melanoleucas* and *Arizona elegans*, and through the ventral scutes into the body cavity in *Lampropeltis getulus* and *Elaphe guttata*. This effort was generally successful with little movement (seven out of 17 reported). Connective tissue formed around most of the tags, securing them in position.

The impact of PIT tagging on the speed and growth rate of neonatal *Thamnophis marcianus* was studied by Keck (1994). Tags (125 kHz model 1400L, Biosonics Corp., Seattle, Washington, USA) were injected caudally to the stomach and cranial to the gonads in snakes, seven days postparturition, using a 12-guage needle. No significant difference was found in growth, speed or survival between the control and experimental groups.

In a similar study, Jemison et al. (1995) found PIT tagging in *Sistrurus miliarius* to have no significant impact on growth and movement in snakes more than one year old. The authors suggest that intraabdominal injection increases chances of tag loss through the urinary or digestive systems and increase the risk of internal injury over intramuscular or subcutaneous injections. Superficial tag placement, however, may increase the chance of tag damage or loss through the skin (Jemison et al., 1995).

RADIO TRANSMITTER TAGGING

The use of radio transmitters as a marking device may be applicable to certain studies. Often these devices are also used to monitor and transmit physiological parameters (temperature, heart rate, etc.). As a marking technique, radio transmission offers such advantages as providing a constant record of the animal's location (day or night, above or below substrate) and allowing the investigator to remain some distance from the animal and, therefore, have a minimal influence on its behavior. Disadvantages of this technique include its relative complexity, cost, feasibility depending on the size of the animal, and limitations of the power source. In this review I will attempt to indicate how radio telemetry has been used with amphibians and reptiles and hopefully give enough information so that you can determine if this technique is appropriate for your study. Due to the complexity of the technique, however, an investigator is expected to consult other sources in adapting this procedure to a specific problem. For example, surgical implantation is required in many applications which may involve extensive training and background.

Many investigators construct their own transmitters or choose from the many that are now commercially available. Suppliers used in specific projects will be mentioned below. If you use radio transmission and have a limited background in electronics, whether or not you construct your own system, reading the chapter on electronics in Mackay (1970) and the review by Kenward (1987) is highly recommended.

The means of receiving and converting the transmitted signals must be precisely selected. The receiver and antenna system should be capable of detecting very weak signals which is somewhat limited by the noise arising from the receiving equipment and the environment. The signal-to-noise ratio increases with the use of a narrow receiver if the transmitter frequency falls within the receiver passband. This condition, in turn, demands a known and stable transmitter frequency and precise receiver tuning. Once the signal is detected, the bearing from which it is coming requires some form of directional receiving antenna to allow the locating of the animal (Shirer and Downhower, 1968). A good description of basic equipment is provided by Kenward (1987).

FROGS AND TOADS

In amphibians radio transmitters are either implanted into the coelomic cavity or fastened externally using a harness. Gray et al. (2005) documented the impact and reception quality of implanted transmitters on

Bufo cognatus. SM1-H transmitters (AVM Instrument Company) weighing 6.63 g and 2.45 cm^3 in volume were implanted in toads anesthetized with a 1000 mg/L solution of tricaine methanesulfonate which took from 10 to 32 minutes. After making a 20 mm incision in the skin of the left caudoventrolateral quadrant of the abdomen with a #20 sterile scalpel, the abdominal muscle was punctured with an surgical scissors and the opening separated for the insertion of the transmitter. Both layers were then sutured and the skin surface dried and sealed with surgical glue after the 8 to 10 minute procedure. Recovery was in a 1% solution of antibacterial tetracaine powder and took about 12 minutes. The receiver used by Gray et al. (2005) was the R2100 Advanced Telemetry Systems model with a 3-element AF Antronics Yagi antenna (model #F152-3FB). One of the ten toads used in this lab study died due to complications from surgery where a suture pulled through the skin and resulted in a secondary infection. All transmitters were eventually encapsulated by connective tissue within the body cavity with no known adverse effects as determined by necropsy. As expected, above ground signal reception decreased linearly throughout the 4 month life span of the transmitter battery.

Watson et al. (2003) monitored the movements of adult *Rana pretiosa* using VHF transmitters (Holohil Systems, Ltd.) attached with nylon ribbon waistbands. The lighter weight males were fitted with BD-2 transmitters (0.9-1.2 g) and the heavier females with BD-2G transmitters (1.2-2.0 g) having 7 and 14 week expected battery life respectively. A later report on the same study (McAllister et al., 2004) indicated that for 94 attempts transmitter use was terminated due to belts slipping off (34%), battery failure (31%), transmitter failure (23%) and antenna entanglement (1%) over an average monitoring period of 57 days.

A variety of waistband attachments for anurans have been used. An aluminum ball or beaded chain belt was developed for use on *Rana aurora* by Rathbun and Murphey (1996) as shown in Figure 16 (size #3, Ball Chain Manufacturing co., 471 South Fulton Ave., Mt. Vernon, New York 10550, USA or Bead Industries, 110 Mt. Grove St., Bridgeport, Connecticut 06605, USA). The transmitters fastened the belt with Devcon two-ton epoxy were model BD-2G (Holohil Systems Ltd., 112 John Cavanagh Rd., Carp, Ontario, Canada KOA 1L0). While some loss and impact of these belts was noticed, the authors believe them to be an improvement over other attachments available at the time. Bull (2000) compared attachments on *Rana luteiventris* with one being a 6 mm satin ribbon with the ends stitched together with carpet thread just tight enough to prevent loss over the hind legs. The transmitter was fixed to this ribbon with cyanoacrylate superglue. The second mode of attachment used by Bull (2000) was with carpet thread tied in a square knot around the upper arm (see Figure 17). The author concluded that the arm band was a better choice with gravid females where the waist ribbon seemed to interfere with oviposition, but the thread of the arm bands was abrasive and is best used for periods no longer than one week.

Radio transmitters were attached to *Bufo boreas* by Bartelt and Peterson (2000) using a waist band of surgical grade microcatheter (polyethylene) tubing (ID = 0.58 mm, OD = 0.965 mm from VMR Scientific). The ends were secured with a large-size flyline eyelet (Al's Goldfish Lure Co.,

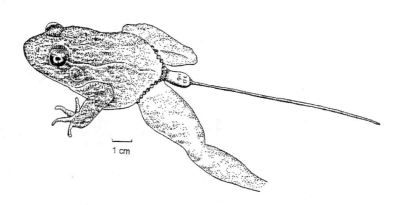

Figure 16. An aluminum ball or beaded chain belt was developed for use on *Rana aurora* by Rathbun and Murphey (1996).

Figure 17. Bull (2000) attached a transmitter with carpet thread tied in a square knot around the upper arm of a Columbia spotted frog.

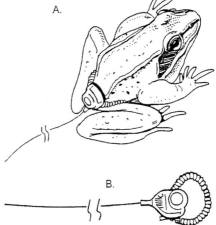

Figure 18. A. Male wood frog with transmitter and belt assembly (from Muths, 2003).
 B. Sketch of belt assembly with radio transmitter for small frogs.

P.O. Box 51013, Indian Orchard, MA 01051, USA) secured with superglue. The belt was passed through a hole in a model BD-2GT transmitter with a temperature sensitive thermistor (Holohil Systems, Ltd.; Carp, Ontario, Canada). No adverse effects of the transmitters on the behavior of the toads was observed. Nine of 38 tagged toads lost their belts within two weeks of attachment.

Yet another radio transmitter belt was used on *Rana pipiens* and *R. sylvatica* by Muths (2003). Holohil BD-2A transmitters (0.61 g) were designed so the battery was on the top rather than on the front which made it shorter and easier for frogs to orient themselves in the water (see Figure 18). Fine craft elastic (gossamer floss, B. Toucan, Inc., USA at $1.64/5yds) and Japanese glass seed beads (size 14; F 458, US $3.45 for thousands) in olive matte were used to make the belt which was stretched over the extended hind legs to rest around the inguinal area (Figure 18). The whip antenna of the transmitter was shortened to 9.5 cm to minimize chances of entanglement. Muths (2003) found the advantages to this system to be its light weight, ability to blend color to the animal, low cost and flexibility in sizing. In addition it appears that the belt will tear away if the belt does become entangled (six of 13 were shed in the study). No problems with abrasion were detected over a three-week period. Adjusting the belt size to the individual frogs in the field can be difficult in cold temperatures due to the need to use tweezers or a fine needle in threading the beads.

Weick et al. (2005) evaluated four methods of external attachment for transmitters (nickel bead chains, aluminum bead chains, plastic cable ties and sewn elastic bands) and two methods of internal surgical implantation (subcutaneous and peritoneal) in *Rana pipiens*. The aluminum bead chain and sewn elastic had the least loss and skin lesions of the external methods, but all these techniques "proved inadequate for tracking *R. pipiens* over long time periods because frogs shed their belts quickly or developed lesions." Peritoneal implants (Figure 19) were more successful in tracking frogs over a

Figure 19. Incision (arrow) from surgically implanted transmitter in *Rana pipiens*. (Photo courtesy of USGS in Weick, et al., 2005).

several month study even though nonrecovery (N = 17 of 90) from anesthesia was an issue. The authors conclude that all techniques tested have "significant health risks for frogs". It is recommended that techniques be chosen carefully, skills be developed through practice with test animals and input from experienced investigator and surgeries be conducted under closely controlled conditions.

TURTLES

Legler (1979) provides a good starting point for anyone interested in using radio telemetry with turtles. In a mark-recapture study of *Chelydra serpentina* by Obbard and Brooks (1981) homemade 150 MHz band transmitters measuring 3 X 10 cm, 200 g and with 30 cm whip antennae were attached to the marginal scutes on the carapace with brass bolts. These tags weighed less than 5% of the body weight of the smallest turtle and lasted (using two Mallory Rm-4R mercury cells) up to 8 months (Obbard and Brooks, 1981). Radio transmitters (AVM Instrument Co. or Custom Electronics) used to track movements of *Chrysemys picta* by Rowe (2003) were sealed with a plastic coat (Plastidip ®). The package was then secured to the carapace on the dorsal rear margin with wire through two holes drilled in the marginal scutes and then sealed with epoxy. The antenna was fastened around the marginal scutes on the dorsal surface with epoxy bringing the total weight of the transmitter to 8.4 g or about 2.4 % of the average turtle body mass (Rowe, 2003).

Boarman et al. (1998) provide an extensive review (113 accounts) of attachments techniques for radio transmitters on turtles including use of adhesives, harnesses, holes in the shell, implantation and tape. They also report on three techniques they developed for use of transmitters on *Gopherus agassizii* depending on the size of the tortoise (see Figure 20). The three transmitters used were: 1) A 35 g two-stage battery-powered model (AVM Instruments SB-2) for tortoises (N = 43) with a 171-296 mm MCL (midline carapace length), 2) A 26 g one-stage battery powered model (AVM Instruments SM-1H) for tortoises (N = 14) between 146 and 239 mm MCL, and 3) A 4.2 g one-stage solar-assisted model (AVM Instruments SM-1H-solar) for immature and subadult tortoises (N = 21) between 97 and 207 mm MCL. The whip antennae for the two non-solar transmitters were made of 20 gauge, insulated, stranded wire ranging from 280 to 320 mm. Those for the solar transmitters were 24 gauge and 150 mm long. Additional details on use of adhesives for transmitter attachment are given by Boarman et al. (1998) and study of these is recommend before using their technique.

LIZARDS

The use of telemetry with large lizards is made somewhat less difficult due to the ability to attach the transmitters externally. Montgomery et al. (1973) packaged the transmitter, battery, battery leads and antennae to form a harness placed around the chest and neck of *Iguana iguana*. A similar

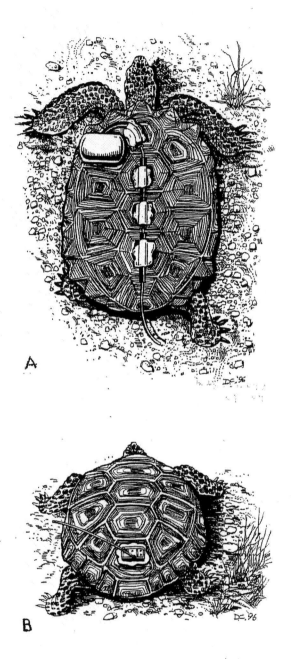

Figure 20. Boarman et al. (1998) attached radio transmitters to carapace of desert tortoises.
    A.   Larger, battery-powered transmitter weighing 800g.
    B.   Smaller, solar-assisted transmitter weighing 220 to 1800 g.

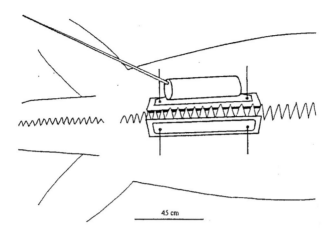

Figure 21. Illustration from Goodman (2005) of the method used to suture a transmitter below the dorsal crest of an iguana.

backpack harness was fashioned for use on tuatara (*Sphenodon punctatus*) by Ussher (1999). In a study of *Cyclura lewisi*, Goodman (2005) reported that suturing transmitters below the dorsal crest (see Figure 21) had a much better retention rate than transmitters which were glued to the posterior dorsum. Attachment of transmitters to *Cyclura cyclura* using 20 lb. test monofilament line by Knapp and Owens (2005) worked well with the caution that young individuals need to be recaptured regularly to monitor abrasion and to enlarge the lines to adjust for growth.

As with the other categories of amphibians and reptiles, implantation of transmitters in lizards requires careful development of an appropriate technique for each species. If possible, finding another investigator who has used a procedure similar to what seems best for your animal would be the most efficient way to proceed.

SNAKES

As with lizards, the larger the snake, the more easily it can carry a transmitter. Some companies manufacture transmitters specifically for various sizes of snakes, large enough to provide a good transmission range, but as small as possible to easy implantation and disturbance of behavior. Batteries too must balance size with ability to broadcast an adequate signal while minimizing the impact on the biology of the snake. The transmitter/battery package is often coated with silicone rubber to protect it from body fluids and streamline the shape for surgical insertion or force feeding. Either of these methods of placing the transmitter package in the snake can be quite disruptive to the snake. (Fitch, 1987).

Movements of eight species of snakes in Kansas were studied using force-fed transmitters for up to about one month at a time (Fitch and Shirer, 1971). The authors indicated that force fed transmitters seem to slow down snake activity as hunting/foraging behavior is suppressed. Similar techniques were reported used with *Coluber constrictor* by Brown and Parker (1976a) and *Helicops angulatus* and *Bothrops atrox* by Henderson et al. (1976). Radio telemetry is also used to study physiological activity such as thermoregulation as done with *Natrix fasciata* and *N. taxispilota* by Osgood (1970). A detailed account and extensive illustrations for surgical implantation of radio transmitters including antenna in snakes (*Crotalus horridus* and *Agkistrodon contortrix*) is provided by Reinert and Cundall (1982). Another advancement in the implantation technique was described by Weatherhead and Anderka (1984) for 15 g, 12.5 X 52 mm transmitters and battery with 20 cm antenna in adult *Elaphe obsoleta*. A similar procedure for implantation in *Crotalus viridis* (Stark, 1986) provides some additional information useful for those developing this technique for use with other species. Attachment of small and light weight transmitters (Holohil Systems, Ltd., Model LB-2N) to the dorsal surface of the tail was successful in tracking *Crotalus viridis* to their hibernacula about 50% of the time (Hallock, 2006). Hallock points out that this technique precludes any need for surgical implantation and exposure of the snakes to anesthesia and infection. Lang (1992) provides a comprehensive review of marking techniques for snakes including radiotelemetry.

A recent study of *Lichanura trivirgata* by Diffendorfer et al. (2005) used TRX-1000S (Wildlife Materials, Inc.) , LA 12-Q (AVM Instrument Co.) and TR-4 and TR-5 (Telonics, Inc.) receivers with RA-2AK (Telonics, Inc.) directional antennae. Keeping transmitters at weights less than 5% of the body mass, Holohil Systems Ltd. Model SB-2 was used with smaller snakes and model SI-2T was used with larger individuals. Surgical techniques were performed by veterinarians following procedures similar to those in the publications mentioned above.

CROCODILIANS

Kay (2004) reviews techniques for attaching radio transmitters to crocodilians with description of new method used on *Crocodylus porosus* which can also be adapted for time-depth recorders, satellite tags or GPS data loggers. Crocodiles were physically restrained and blindfolded without anesthesia and tags attached to the large nuchal scales of the dorsum of the neck (Figure 22). These scales were chosen for their raised keels and ability to receive bone pins through the aluminum bracket to which the tag was attached. For attachment the nuchal scales were cleaned with a chlorhexidene scrub, rinsed with water, dried, and sprayed with 70% EtOH. When dry a lump of glue (either Loctite Fixmaster Underwater Repair Epoxy, www.loctite.com or Selleys Knead It Aqua, www.selleys.com.au) was used to place the tag between the scales, molding it to reduce the profile of the tag

(Fig. 22). Kirschner wires (stainless steel K-wires) were used as bone pins (31 cm X 1.6 mm diameter) and drilled directly through the keel of the scales with the protruding ends then bent with pliers and trimmed with wire cutters. Glue was also used to encase the pins and lower half of the tag to provide a smooth contour.

The bracket technique was successfully used with ten tags, but since the glue seemed to be holding so well, an additional six crocodiles were tagged with no brackets (Kay, 2004). Of the 16 tags, three detached in snags, one detached when escaping a trap, and one was intentionally removed to determine any negative impact on the integument. The glue only modification was just as successfully as use of brackets and saved considerable time and effort in the tagging operation. The author suggests that alternative bone pin choices might be titanium or Delrin plastic (www.plastics.dupont.com) as being more inert than stainless steel and Sikaflex-291 (www.sika-industry.com) flexible polyurethane glue as being less dense, but requiring a long curing time.

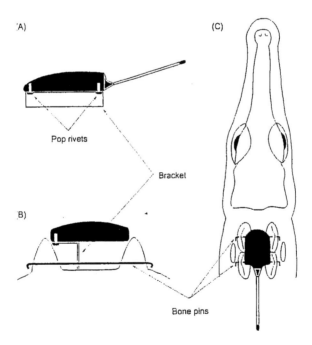

Figure 22. Kay (2004) illustrated the placement and orientation of the radio transmitter, bracket and bone pins on nuchal scales of a crocodilian. Attachment was facilitated with glue, which so well to the tag that bone pins holding the bracket may now be redundant.

LITERATURE CITED

Alvarado, J., A. Figueroa, C. Delgado, M. T. Sanchez, and E. Lopez. 1993. Differential retention of metal and plastic tags on the black sea turtle (*Chelonia agassizi*). Herpetological Review 24:23-24.

Anholt, B. R., S. Negovetic, and C. Som. 1998. Methods for anaesthetizing and marking larval anurans. Herpetological Review 29:153-154.

Arntzen, J. W., A. Smithson, and R. S. Oldham. 1999. Marking and tissue sampling effects on body condition and survival in the newt *Triturus cristatus*. Journal of Herpetology 33:567-576.

Ashton, D. G. 1978. Marking zoo animals for identification. *In*: Stonehouse, B. (ed.). Animal Marking: Recognition Marking of Animals in Research. University Park Press, Baltimore, Maryland.

Ashton, R. W. 1975. A study of movement, home range, and winter behavior of *Desmognatus fuscus* (Rafinesque). Journal of Herpetology 9:85-91.

Bacon, P. R. 1973. The orientation circle in the beach ascent crawl of the leatherback turtle, *Dermochelys coriacea*, in Trinidad. Herpetologica 29:343-348.

Bailey, L. L. 2004. Evaluating elastomer marking and photo identification methods for terrestrial salamanders: marking effects and observer bias. Herpetological Review 35:38-41.

Balazs, G. H. 1985. Retention of flipper tags on hatchling sea turtles. Herpetological Review 16:43-45.

Barbour, R. W., J. W. Hardin, J. P. Shafer, and M. J. Harvey. 1969. Home range, movements, and activity of the dusky salamander, *Desmognathus fuscus*. Copeia 1969:293-287.

Bartelt, P. E., and C. R. Peterson. 2000. A description and evaluation of a plastic belt for attaching radio transmitters to western toads (*Bufo boreas*). Northwestern Naturalist 81:122-128.

Bayless, L. E. 1975. Population parameters for *Chrysemys picta* in a New York pond. American Midland Naturalist 93:168-176.

Bennett, D. H., J. W. Gibbons, and J. C. Franson. 1970. Terrestrial activity in aquatic turtles. Ecology 51:738-740.

Bennion, R. S., and W. S. Parker. 1976. Field observations on courtship and aggressive behavior in desert striped whipsnakes, *Masticophis t. taeniatus*. Herpetologica 32:30-35.

Bjorndal, K. A., A. B. Bolten, C. J. Lagueux, and A. Chaves. 1996. Probability of tag loss in green turtles nesting at Tortuguero, Costa Rica. Journal of Herpetology 30:567-571.

Blanchard, F. N., and E. B. Finster. 1933. A method of marking living snakes for future recognition with a discussion of some problems and results. Ecology 14:334-347.

Blankenship, E. L., T. W. Bryan, and S. P. Jacobson. 1990. A method for tracking tortoises using fluorescent powder. Herpetological Review 21:88-89.

Bloch, N., and D. J. Irschick. 2005. Toe-clipping dramatically reduces clinging performance in a pad-bearing lizard (*Anolis carolinensis*). Journal of Herpetology 39:293-298.

Boarman, W. I., T. Goodlett, G. Goodlett, and P. Hamilton. 1998. Review of radio transmitter attachment techniques for turtle research and recommendations for improvement. Herpetological Review 29:26-33.

Boone, J., and E. Larue. 1999. Effects of marking *Uta stansburiana* (Sauria: Phrynosomatidae) with xilene-based paint. Herpetological Review 30:33-34.

Borges-Landaez, P. A., and R. Shine. 2003. Influence of toe-clipping on running speed in *Eulamprus quoyii*, an Australian scincid lizard. Journal of Herpetology 37:592-595.

Bradfield, K. S. 2004. Photographic identification of individual Archey's frogs, *Leiopelma archeyi*, from natural markings. DOC Science Internal Series 191, New Zealand Department of Conservation, Wellington.

Breckenridge, W. J., and J. R. Tester. 1961. Growth, local movements and hibernation of the Manitoba toad, *Bufo hemiophrys*. Ecology 42:637-646.

Briggs, J. L., and R. M. Storm. 1970. Growth and population structure of the cascade frog, *Rana cascadae* Slater. Herpetologica 26:283-300.

Broderick, A. C., and B. J. Godley. 1999. Effect of tagging marine turtles on nesting behaviour and reproductive success. Animal Behavior 58:587-591.

Brown, L. J. 1997. An evaluation of some marking and trapping techniques currently used in the study of anuran population dynamics. Journal of Herpetology 31:410-419.

Brown, W. C., and A. C. Alcala. 1970. Population ecology of the frog, *Rana erythraea* in southern Negros, Philippines. Copeia 1970:611-622.

Brown, W. S., and W. S. Parker. 1976a. Movement ecology of *Coluber constrictor* near communal hibernacula. Copeia 1976:225-242.

Brown, W. S., and W. S. Parker. 1976b. A ventral scale clipping system for permanently marking snakes (Reptilia, Serpentes). Journal of Herpetology 10:247-249.

Brown, W. S., V. P. J. Ganon, and D. M. Secoy. 1984. Paint-marking the rattle of rattlesnakes. Herpetological Review 15:75-76.

Buchan, A., L. Sun, and R. S. Wagner. 2005. Using alpha numeric fluorescent tags for individual identification of amphibians. Herpetological Review 36:43-44.

Buhlmann, K. A., and T. D. Tuberville. 1998. Use of Passive Integrated Transponder (PIT) tags for ,marking small freshwater turtles. Chelonian Conservation Biology 3:102-104.

Bull, E. L. 2000. Comparison of two radio transmitter attachments on Columbia spotted frogs (*Rana luteiventris*). Herpetological Review 31:26-28.

Bull, E. L., R. Wallace, and D. H. Bennett. 1983. Freeze-branding: a long term marking technique on long-toed salamanders. Herpetological Review 14:81-82.

Burger, J. 1976. Temperature relationships in nests of the northern diamondback terrapin *Malaclemys terrapin*. Herpetologica 32:412-418.

Burger, J., and W. A. Montevecchi. 1975. Nest site selection in the terrapin, *Malaclemys terrapin*. Copeia 1975:113-119.

Burke, R. L., C. M. Schneider, and M. T. Delinger. 2005. Cues used by raccoons to find turtle nests: effects of flags, human scent, and diamond-backed terrapin sign. Journal of Herpetology 39:312-315.

Butler, B. O., and T. E. Graham. 1993. Tracking hatchling Blanding's turtles with fluorescent pigments. Herpetological Review 24:21-22.

Cagle, F. R. 1939. A system of marking turtles for future identification. Copeia 1939:170-172.

Camper, J. D., and J. R. Dixon. 1988. Evaluation of a microchip marking system for amphibians and reptiles. Texas Parks and Wildlife Department, Research Publication 7100-159:1-22.

Carlstrom, D., and E. Edelstam. 1946. Methods of marking reptiles for identification after recapture. Nature 158:748-749.

Carpenter, C. C. 1954. A study of amphibian movement in the Jackson Hole Wildlife Park. Copeia 1954:197-200.

Carr, A. 1967. So Excellente A Fish: An Natural History of Sea Turtles. Natural History Press, New York.

Carr, A., P. Ross, and S. Carr. 1974. Internesting behavior of the green turtle, *Chelonia mydas*, at a mid-ocean island breeding ground. Copeia 1974:703-706.

Cecil, S. G., and J. J. Just. 1978. Use of acrylic polymers for marking tadpoles (Amphibia, Anura). Journal of Herpetology. 12:95-96.

Chabreck, R. H. 1965. Methods of capturing, marking and sexing alligators. Proceedings 17th Annual Conference Southeastern Association of the Game and Fish Commission. Pp. 47-50.

Clark, D. L., and J. C. Gillingham. 1984. A method for nocturnally locating lizards. Herpetological Review 15:24-25

Clark, D. R. 1971. Branding as a marking technique for amphibians and reptiles. Copeia 1971:148-151.

Clarke, R. D. 1972. The effect of toe clipping on survival in Fowler's toad (*Bufo woodhousei fowleri*). Copeia 1972:182-185.

Cowin, S., and J. Cebek. 2006. Feasibility of using plastron markings in young wood turtles (*Glyptemys insculpta*) as a technique for identifying individuals. Herpetological Review 37:305- 307.

Daniel, J. A., K. A. Baker, and K. E. Bonine. 2006. Retention rates of surface and implantable marking methods in the Mediterranean house gecko (*Hemidactylus turcicus*), with notes on capture methods and rates of skin shedding. Herpetological Review 37:319-321.

Daugherty, C. H. 1976. Freeze branding as a technique for marking anurans. Copeia 1976:836-838.

Davis, T. M., and K. Ovaska. 2001. Individual recognition of amphibians: effects of toe clipping and fluorescent tagging on the salamander *Plethodon vehiculum*. Journal of Herpetology 35:217-225.

Davis, W., and G. Sartor. 1975. A method of observing movements of aquatic turtles. Herpetological Review 6:13-14.

Deavers, D. R. 1972. Water and electrolyte metabolism in the arenicolous lizard *Uma notata notata*. Copeia 1972:109-122.

Diffendorfer, J. E., C. Rochester, R. N. Fisher, and T. K. Brown. 2005. Movement and space use by coastal rosy boas (*Lichanura trivirgata roseofusca*), in coastal southern California. Journal of Herpetology 39:24-36.

Dodd, C. K. 1993. The effects of toe-clipping on sprint performance of the lizard *Cnemidophorus sexlineatus*. Journal of Herpetology 27:209-213.

Dole, J. W. 1965. Summer movements of adult leopard frogs, *Rana pipiens* Schreber, in northern Michigan. Ecology 46:236-255.

Donnelly, M. A., C. Guyer, J. E. Juterbock, and R. A. Alford. 1994. Techniques for marking amphibians. *In*: Heyer, W. R., M. A. Donnelly, R. W. McDiarmid, L. C. Hayek, and M. S. Foster (eds), Measuring and Monitoring Biological Diversity, Standard Methods for Amphibians, Appendix 2, pp. 277-284. Smithsonian Institution Press, Washington.

Doody, J. S. 1995. A photographic mark-recapture method for patterned amphibians. Herpetological Review 26:19-21.

Dreslik, M. J. 1997. An inexpensive method for creating spaghetti tags for marking Trionychid turtles. Herpetological Review 28:33.

Eckert, K. L., and S. A. Eckert. 1989. The application of plastic tags to leatherback sea turtles *Dermochelys coriacea*. Herpetological Review 20:90-91.

Efford, I. E., and J. A. Mathias. 1969. A comparison of two salamander populations in Marian Lake, British Columbia. Copeia 1969:723-736.

Elmberg, J. 1989. Knee-tagging—a new marking technique for anurans. Amphibian-Reptilia 10:101-104.

Emlen, S. T. 1968. A technique for marking anuran amphibians for behavioral studies. Herpetologica 24:172-173.

Ernst, C. H. 1971. Population dynamics and activity cycles of *Chrysemys picta* in southeastern Pennsylvania. Journal of Herpetology 5:151-160.

Ernst, C. H., M. F. Hershey, and R. W. Barbour. 1974. A new coding system for hardshelled turtles. Transactions of the Kentucky Academy of Science 35:27-28.

Fellers, G. M., and C. A. Drost. 1989. Fluorescent powder-a method for tracking reptiles. Herpetological Review 20:91-92,

Ferner, J. W. 1979. A review of marking techniques for amphibians and reptiles. Society for the Study of Amphibians and Reptiles, Herpetological Circular No. 9.

Fisher, M., and A. Muth. 1989. A technique for permanently marking lizards. Herpetological Review 20:45-46.

Fitch, H. S. 1987. Collecting and life-history techniques. Pp. 143-164. *In*: Seigel, R. A., J. T. Collins, and S. S. Novak (eds). Snakes: ecology and evolutionary biology. Macmillan Publishing Company, New York.

Fitch, H. S., and H. W. Shirer. 1971. A radiotelemetric study of spatial relationships in some common snakes. Copeia 1971:118-128.

Fontaine, C. T., T. A. Williams, and J. D. Camper. 1987. Ridleys tagged with passive integrated transponders (PIT). Marine Turtle Newsletter 41:6.

Froese, A. D., and G. M. Burghardt. 1975. A dense natural population of the common snapping turtle (*Chelydra s. serpentina*). Herpetologica 31:204-208.

Funk, W. C., M. A. Donnelly and K. R. Lips. 2005. Alternative views of amphibian toe-clipping. Nature 433:193.

Galbraith, D. A., and R. J. Brooks. 1984. A tagging method for use in hatchling turtles. Herpetological Review 15: 73, 75.

Gaymer, R. 1973. A marking technique for giant tortoises and field trials in Aldabra. Journal of Zoology, London 169:393-401.

George, I. D. 1940. Marking of frogs for future reference in natural history studies. Copeia 1940:134.

Germano, D. J., and D. F. Williams. 1993. Field evaluation of using Passive Integrated Transponder (PIT) tags to permanently mark lizards. Herpetological Review 24:54-56.

Goodman, R. M. 2005. Attachment of radio transmitters in a rock iguana, *Cyclura lewisi*. Herpetological Review 36:150-152.

Gower, D. J., O. V. Oommen, and M. Wilkinson. 2006. Marking amphibians with alpha numeric fluorescent tags: caecilians lead the way. Herpetological Review 37:302.

Graham, T. E. 1986. A warning against the use of Peterson disc tags in turtle studies. Herpetological Review 17:42-43.

Grant, E.H. C. and P. Nanjappa. 2006. Addressing error in identification of *Ambystoma maculatum* (Spotted Salamanders) using spot patterns. Herpetological Review 37(1):57-60.

Gray, M. J., D. L. Miller, and L. M. Smith. 2005. Coelomic response and signal range of implant transmitters in *Bufo cognatus*. Herpetological Review 36: 285-288.

Grubb, J. C. 1970. Orientation in post-reproductive Mexican toads, *Bufo valliceps*. Copeia 1970:674-680.

Guttman, S. I., and W. Creasy. 1973. Staining as a technique for marking tadpoles. Journal of Herpetology 7:388-390.

Hagstrom, T. 1973. Identification of newt specimens (Urodela, *Triturus*) by recording the belly pattern and a description of photographic equipment for such registrations. British Journal of Herpetology 4:321-326.

Hall, R. J., and D. P. Stafford. 1972. Studies in the life history of Wehrles salamander, *Plethodon wehrlei*. Herpetologica 28:300-309.

Halliday, T. 1994. Marking amphibians by toe-clipping. Froglog 10:2-3.

Halliday, T. 1995. More on toe-clipping. Froglog 12:2-3.

Hallock, L. A. 2006. External attachment of radio-transmitters on western rattlesnakes (*Crotalusviridis*) to locate communal hibernacula. [Abstract]. Northwestern Naturalist 87(2):170.

Healy, W. R. 1974. Population consequences of alternative life histories in *Notophthalmus v.viridescens*. Copeia 1974:221-229.

Healy, W. R. 1975. Terrestrial activity and home range in efts of *Notophthalmus viridescens*. Copeia 1975?: 221-229.?

Heatwole, H. 1961. Inhibition of digital regeneration in salamanders and its use in marking individuals for field studies. Ecology 42:593-594.

Henderson, R. W. 1974. Aspects of the ecology of the juvenile common iguana (*Iguana iguana*). Herpetologica 30:327-332.

Henderson, R. W., M. A. Nickerson, and S. Ketcham. 1976. Short term movements of the snakes *Chironius carinatus*, *Helicops angulatus* and *Bothrops atrox* in Amazonian Peru. Herpetologica 32:304-310.

Hendrickson, J. R. 1954. Ecology and systematics of salamanders of the genus *Batrochoseps*. University of California Publications in Zoology 54:1-46.

Henley, G. B. 1981. A new technique for recognition of snakes. Herpetological Review 12:56.
Henwood, T. A. 1986. Losses of monel flipper tags from loggerhead seas turtles, *Caretta caretta*. Journal of Herpetology 20:276-279.
Hero, J. M. 1989. A simple code for toe clipping anurans. Herpetological Review 20:66-67.
Herreid, C. F., and S. Kinney. 1966. Survival of Alaskan woodfrog (*Rana sylvatica*) larvae. Ecology 47:1039-1041.
Hillis, R. E., and E. D. Bellis. 1971. Some aspects of the ecology of the hellbender, *Cryptobranchus alleganiensis alleganiensis*, in a Pennsylvania stream. Journal of Herpetology 5:121-126.
Hirth, H. G. 1966. Weight changes and mortality in three species of snakes during hibernation. Herpetologica 22:8-12.
Hudnall, J. A. 1982. New techniques for measuring and tagging snakes. Herpetological Review 13:97-98.
Hudson, S. 1996. Natural toe loss in southeastern Australian skinks: implications for marking lizards by toe-clipping. Journal of Herpetology 30:106-110.
Ireland, D., N. Osbourne, and M. Berrill. 2003. Marking medium to large sized anurans with Passive Integrated Transponder (PIT) tags. Herpetological Review 34:218-220.
Ireland, P. H. 1973. Marking larval salamanders with fluorescent pigments. Southwestern Naturalist 18:252-253.
Jameson, D. L. 1957. Population structure and homing responses in the Pacific tree frog. Copeia 1957:221-228.
Jemison, S. C., L. A. Bishop, P. G. May, and T. M. Farrell. 1995. The impact of PIT-tags on growth and movement of the rattlesnake, *Sistrurus miliarius*. Journal of Herpetology 29:129-132.
Jennings, M. L., D. N. David, and K. M. Portier. 1991. Effect of marking techniques on growth and survivorship of hatchling alligators. Wildlife Society Bulletin 19:204-207.
Jenssen, T. A. 1970. The ethoecology of *Anolis nebulosus* (Sauria, Iguanidae). Journal of Herpetology 4:1-38.
Joanen, T. and L. McNease. 1973. A telemetric study of adult male alligators of Rockefeller Refuge, Louisiana. Proceedings of the 26th Annual Conference of Southeastern Association Game and Fish Commissions, pp. 252-275.
Johnson, M. A. 2005. A new method of temporarily marking lizards. Herpetological Review 36(3): 277-279.
Jones, S. M., and G. W. Ferguson. 1980. The effect of paint marking on mortality in a Texas population of *Sceloporus undulatus*. Copeia 1980:850-854.
Judd, F. W. 1975. Activity and thermal ecology of the keeled earless lizard, *Holbrookia propinqua*. Herpetologica 31:137-150.
Kaplan, H. M. 1958. Marking and banding frogs and turtles. Herpetologica 14:131-132.
Karlstrom, E. L. 1957. The use of Co60 as a tag for recovering amphibians in the field. Ecology 38:187-195.

Kay, W. R. 2004. A new method for attaching electronic devices to crocodilians. Herpetological Review 35:354-357.

Keck, M. B. 1994. Test for detrimental effects of PIT tags in neonatal snakes. Copeia 1994:226-228.

Keller, C. 1993. Use of fluorescent pigment for tortoise nest location. Herpetological Review 24:140-141.

Kenward, R. 1987. Wildlife Radio Tagging: Equipment, Field Techniques and Data Analysis. Academic Press, London. 222 pp.

Kinkead, K. E., J. D. Lanham and R. R. Montanucci. 2006. Comparison of anesthesia and marking techniques on stress and behavioral responses in two *Desmognatus* salamanders. Journal of Herpetology 40:323-328.

Knapp, C. R., and A. K. Owens. 2005. An effective new radio transmitter attachment technique for lizards. Herpetological Review 36:264-266.

Kurashina, N., T. Utsunomiya, Y. Utsunomiya, S. Okada, and I. Okochi. 2003. Estimating the population size of an endangered population of *Rana porosa brevipoda* Ito (Amphibian: Ranidae) from photographic identification. Herpetological Review 34:348-349.

Lackey, D., M. Matheson, L. Sheeran, L. Jinhua and R. S. Wagner. 2006. Demography and non-invasive individual identification using spot patterns in Chinese salamanders (*Pachytriton brevipes*). [Abstract]. Northwestern Naturalist 87(2):176.

Lang, M. 1992. A review of techniques for marking snakes. Smithsonian Herpetological Information Service 90:1-19.

Langkilde, T. and R. Shine. 2006. How much stress do researchers inflict on their study animals? A case study using a scincid lizard, *Eulamprus heatwolei*. Journal of Experimental Biology 209:1035-1043.

Layfield, J. A., D. A. Galbraith, and R. J. Brooks. 1988. A simple method to mark hatchling turtles. Herpetological Review 19:78-79.

LeBuff, C. R. and R. W. Beatty. 1971. Some aspects of nesting of the loggerhead turtle, *Caretta caretta caretta* (Linne) on the Gulf coast of Florida. Herpetologica 27:153-156.

Legler, W. K. 1979. Telemetry. Chapter 3:61-72. In: Harless, M.and H. Morlock (eds). Turtles. Perspectives and Research. Wiley and Sons, New York.

LemKau, P. J. 1970. Movements of the box turtle, *Terrapene c. carolina* (Linnaeus), in unfamiliar territory. Copeia 1970:781-783.

Lewke, R. R., and R. K. Stroud. 1974. Freeze branding as a method of marking snakes. Copeia 1974:997-1000.

Loafman, P. 1991. Identifying individual spotted salamanders by spot pattern. Herpetological Review 22:91-92.

Lonke, D. J., and M. E. Obbard. 1977. Tag success, dimensions, clutch size and nesting site fidelity for the snapping turtle, *Chelydra serpentina* (Reptilia, Testudines, Chelydridae) in Algonquin Park, Ontario, Canada. Journal of Herpetology 11:243-244.

Lüddecke, H., and A. Amezquita. 1999. Assessment of disc clipping on the survival and behavior of the Andean frog *Hyla labialis*. Copeia 1999:824-830.

Madison, D. H., and C. R. Shoop. 1970. Homing behavior, orientation, and home range of salamanders tagged with Tantalum-182. Science 168:1484-1487.

Martof, B. S. 1953. Territoriality in the green frog, *Rana clamitans*. Ecology 34:165-174.

Mackay, R. S. 1970. Bio-medical Telemetry. New York, John Wiley and Sons, Inc.

May, R. M. 2004. Ethics and amphibians. Nature 431:403.

McAllister, K. R., J. W. Watson, K. Risenhoover, and T. McBride. 2004. Marking and radiotelemetry of Oregon spotted frogs (*Rana pretiosa*). Northwestern Naturalist 85:20-25.

McCarthy, M. A. and K. M. Parris. 2004. Clarifying the effect of toe clipping on frogs with Bayesian statistics. Journal of Applied Ecology 41:780-786.

McDonald, D. L., and P. H. Dutton. 1996. Use of PIT tags and photoidentification to reviseremigration of leatherback turtles (*Dermochelys coriacea*) nesting in St. Croix, U. S. Virgin Islands, 1979-1995. Chelonian Conservation Biology 2:148-152.

McDonald, D., P. Dutton, R. Brandner, and S. Basford. '1996. Use of pineal spot ("pink spot") photographs to identify leatherback turtles. Herpetological Review 27:11-12.

Medica, P. A., G. A. Hoddenbach, and J. R. Lannom. 1971. Lizard sampling techniques. Rock Valley Misc. Publ. No. 1, 55 pp.

Medica, P. A., R. B. Bury, and F. B. Turner. 1975. Growth of the desert tortoise (*Gopherus agassizii*) in Nevada. Copeia 1975:639-643.

Medway, L. 1978. Section 2. Tagging. Pp. 41-42. *In*: B. Stonehouse (ed.). Animal Marking.University Park Press, Baltimore, Maryland.

Minnich, J. E., and V. H. Shoemaker. 1970. Diet, behavior and water turnover in the desert iguana, *Dipsosaurus dorsalis*. American Midland Naturalist 84:496-509.

Montgomery, G. G., A. S. Rand, and M. E. Sunquist. 1973. Post-nesting movements of iguanas from a nesting aggregation. Copeia 1973:620-622.

Moon, B. R., C. S. Ivanyi, and J. Johnson. 2004. Identifying individual rattlesnakes using tail pattern variation. Herpetological Review 35:154-156.

Moosman, D. L., and P. R. Moosman. 2006. Subcutaneous movements of visible implant elastomers in wood frogs (*Rana sylvatica*). Herpetological Review 37:300-301.

Muths, E. 2003. A radio transmitter belt for small ranid frogs. Herpetological Review 34:345-348.

Muths, E., P. S. Corn, and T. R. Stanley. 2000. Use of oxytetracycline in batch-marking post- metamorphic boreal toads. Hereptological Review 31:28-32.

Nace, G. W., and E. K. Manders. 1982. Marking individual amphibians. Journal of Herpetology 16:309-311.

Nickerson, M. A., and C. E. Mays. 1973. A study of the Ozark hellbender, *Cryptobranchus alleganiensis bishopi*. Ecology 54:1164-1165.

Nishikawa, K. C., and P. M. Service. 1988. A fluorescent marking technique for individual recognition of terrestrial salamanders. Journal of Herpetology 22:351-353.

Obbard, M. E., and R. J. Brooks. 1981. A radiotelemetry and mark-recapture study of activity in the common snapping turtle, *Chelydra serpentina*. Copeia 1981:630-637.

Orser, P. N., and D. J. Shine. 1972. Effects of urbanization on the salamander *Desmognathus fuscus fuscus*. Ecology 53:1148-1154.

Osgood, D. W. 1970. Thermoregulation in water snakes studied by telemetry. Copeia 1970:568-571.

Ott, J. A., and D. E. Scott. 1999. Effects of toe-clipping and PIT-tagging on growth and survival in metamorphic *Ambystoma opacum*. Journal of Herpetology 33:344-348.

Parker, W. S. 1976. Population estimates, age structure, and denning habits of whipsnakes, *Masticophis t. taeniatus*, in a norther Utah *Atriplex-Sarcobatus* community. Herpetologica 32:53-57.

Parris, K. M. and M. A. McCarthy. 2001. Identifying effects of toe clipping on anuran return rates: the importance of statistical power. Amphibia-Reptilia 22:275-289.

Paulissen, M. A. 1986. A technique for marking teiid lizards in the field. Herpetological Review 17:16-17.

Paulissen, M. A., and H. A. Meyer. 2000. The effects of toe-clipping on the gecko *Hemidactylus turcicus*. Journal of Herpetology 34:282-285.

Pendlebury, G. B. 1972. Tagging and remote identification of rattlesnakes. Herpetologica 28(4):349-350.

Perera, A., and V. Perez-Mellado. 2004. Photographic identification as a noninvasive marking technique for lacertid lizards. Herpetological Review 35:349-350.

Peterman, W. E. and R. D. Semlitsch. 2006. Effects of tricaine methanesulfonate (MS-222) concentration on anesthetization and recovery in four plethodontid salamanders. Herpetological Review 37:303-304.

Plummer, M. V. 1979. Collecting and marking. Chapter 2:45-60. *In*: Harless, M. and H. Morlock (eds.). Turtles. Perspectives and Research.

Plummer, M. V. and J. W. Ferner. (In Press). Techniques for marking r Reptiles. Appendix II. *In*: R. McDiarmid, C. Guyer, J. W. Gibbons, N. Foster, and N. Chornoff (eds). Reptiles: Measuring and Monitoring Biodiversity. Smithsonian Institution Press, Washington, D. C.

Pough, F. H. 1966. Ecological relationships of rattlesnakes in southeastern Arizona with notes on other species. Copeia 1966:676-683.

Pough, F. H. 1970. A quick method for permanently marking snakes and turtles. Herpetologica 26:428-430.

Pritchard, P. C. H. 1976. Post-nesting movements of marine turtles (Cheloniidae and Dermochelyidae) tagged in the Guianas. Copeia 1976:749-754.

Pyke, G. H. 2005. The use of PIT tags in capture-recapture studies of frogs: a field evaluation. Herpetological Review 36(3):281-285.

Rafinski, J. N. 1977. Autotransplantation as a method for permanent marking of urodele amphibians (Amphibia, Urodela). Journal of Herpetology 11:241-242.

Raney, E. C. 1940. Summer movements of the bullfrog, *Rana catesbeiana* Shaw, as determined by the jaw tag method. American Midland Naturalist 23:733-745.

Raney, E. C. and E. A. Lachner. 1947. Studies on the growth of tagged toads (*Bufo terrestris americanus* Holbrood). Copeia 1947: 113-116.

Rao, M. V. S., and B. S. Rajabai. 1972. Ecological aspects of the agamid lizards, *Sitana ponticeriana* and *Calotes nemoricola* in India. Herpetologica 28:285-289.

Rathbun, G. B., and T. G. Murphey. 1996. Evaluation of a radio-belt for ranid frogs. Herpetological Review 27:187-189.

Reagan, D. P. 1974. Habitat selection in the three-toed turtle, *Terrapene carolina triunguis*. Copeia 1974:512-527.

Reaser, J. 1995. Marking amphibians by toe-clipping: a response to Halliday. Froglog 12:1-2.

Reaser, J. K., and R. E. Dexter. 1996. *Rana pretiosa* (spotted frog). Toe clipping effects. Herpetological Review 27:195-196.

Regester, K. J. and L. B. Woosley. 2005. Marking salamander egg masses with visible fluorescent elastomer: retention time and effect on embryonic development. American Midland Naturalist 153:52-60.

Reinert, H. K., and D. Cundall. 1982. An improved surgical implantation method for radio-tracking snakes. Copeia 1982:702-705.

Rice, T. M., and D. H. Taylor. 1993. A new method for making waistbands for mark anurans. Herpetological Review 24:141-142.

Rice, T. M., S. E. Walker, B. J. Blackstone, and D. H. Taylor. 1998. A new method for marking individual anuran larvae. Herpetological Review 29:92-93.

Richards, C. M., B. M. Carlson, and S. L. Rogers. 1975. Regeneration of digits and forelimbs in the Kenyan reed frog, *Hyperolius viridiflavus ferniquei*. Journal of Morphology 146:431-436.

Riker, W. E. 1956. Uses of marking animals in ecological studies: the marking of fish. Ecology 37:666-670.

Rittenhouse, T. A. G., T. T. Altnether and R. D. Semlitsch. 2006. Fluorescent powder pigments as a harmless tracking method for ambystomids and ranids. Herpetological Review 37:188-191.

Robertson, J. G. M. 1984. A technique for individually marking frogs in behavioral studies. Herpetological Review 15:56-57.

Rodda, G. H., B.C. Bock, G. M. Burghardt, and A. S. Rand. 1988. Techniques for identifying individual lizards at a distance reveal influences of handling. Copeia 1988:905-913.

Rowe, J. W. 2003. Activity and movements of midland painted turtles (*Chrysemys picta martinata*) living in a small marsh system on Beaver Island, Michigan. Journal of Herpetology 37: 342-353.

Rowe, C. L., and S. M. Kelly. 2005. Marking hatchling turtles via intraperitoneal placement of PIT tags: implications for long-term studies. Herpetological Review 36:408-410.

Runyan, A. L., and P. A. Meylan. 2005. PIT tag retention in *Trachemys* and *Pseudemys*. Herpetological Review 36:45-47.
Sajwaj, T. D., S. A. Piepgras and J. W. Lang. 1998. Blanding's Turtle (*Emydoidea blandingii*) at Camp Ripley. Report submitted to the Nongame Wildlife Program, Minnesota Department of Natural Resources, 185 pp.
Schlaepfer, M. A. 1998. Use of a fluorescent marking technique on small terrestrial anurans. Herpetological Review 29:25-26.
Scott, A. F., and J. L. Dobie. 1980. An improved design for a thread trailing device used to study terrestrial movements of turtles. Herpetological Review 11:106-107.
Seale, D., and M. Boraas. 1974. A permanent mark for amphibian larvae. Herpetologica 30:160-162.
Semlitsch, R. D. 1981. Effects of implanted Tanalum-182 wire tags on the mole salamander, *Ambystona talpoideum*. Copeia 1981:735-737.
Shine, C., N. Shine, R. Shine, and D. Slip. 1988. Use of subcaudal scale anomalies as an aid in recognizing individual snakes. Herpetological Review 19:79-80.
Shirer, H. W., and J. F. Downhower, 1968. Radio tracking of dispersing yellow bellied marmots. Transactions of the Kansas Academy of Science 71:463-479.
Shoop, C. R. 1971. A method for short-term marking of amphibians with 24-sodium. Copeia 1971:371.
Simon, C. A., and B. E. Bissinger. 1983. Paint marking lizards: does the color affect survivorship? Journal of Herpetology 17:184-186.
Singh, L. A. K., and H. R. Bustard. 1976. A method to identify individual young gharial (*Gavialis gangeticus*). British Journal of Herpetology 5:669-671.
Smith, C. 1994. Fish tags for observing free-ranging rattlesnakes. Herpetological Review 25:58.
Spellerberg, I. P., and I. Prestt. 1978. Marking snakes. Pp. 133-141. *In*: B. Stonehouse (ed.). Animal Marking. University Park Press, Baltimore, Maryland.
Stamps, J. A. 1973. Displays and social organization in female *Anolis aeneus*. Copeia 1973:264-272.
Stark, M. A. 1984. A quick, easy and permanent tagging technique for rattlesnakes. Herpetological Review 15:110.
Stark, M. A. 1986. Implanting long-range transmitters in prairie rattlesnakes. Herpetological Review 17:17-18.
Stebbins, R. C., and N. W. Cohen. 1973. The effect of parietalectomy on the thyroid and gonads in free-living western fence lizards, *Sceloporus occidentalis*. Copeia 1973:662-668.
Stickel, I. F. 1950. Populations and home range relationships of the box turtle, *Terrapene c. carolina* (Linnaeus). Ecological Monographs 20:351-378.
Stille, W. T. 1950. The loss of jaw tags by toads. Chicago Academy of Science, Natural History Misc. No. 74:1-2.

Stonehouse, B. (ed.). 1978. Animal Marking: Recognition Marking of Animals in Research. University Park Press, Baltimore, Maryland.

Swingland, I. R. 1978. Marking reptiles. Pp. 119-132. *In*: B. Stonehouse (ed). Animal Marking. University Park Press, Baltimore, Maryland.

Taber, C. A., R. F. Wilkinson, and M. S. Topping. 1975. Age and growth of hellbenders in the Niangua River, Missouri. Copeia 1975:633-639.

Taylor, J., and L. Deegan. 1982. A rapid method for mass marking amphibians. Journal of Herpetology 16:172-173.

Thomas, A. E. 1975. Marking anurans with silver nitrate. Herpetological Review 6:12.

Tinkle, D. W. 1958. Experiments with censusing of southern turtle populations. Herpetologica 14:172-175.

Tinkle, D. W. 1967. The life and demography of the side-blotched lizard, *Uta stansburiana*. Misc. Publication, Museum of Zoology, University of Michigan 132:1-182.

Tozetti, A. M., and L. F. Toledo. 2005. Short-term movement and retreat sties of *Leptodactylus labyrinthicus* (Anura: Leptodactylidae) during the breeding season: a spool-and-line tracking study. Journal of Herpetology 39(4):640-644.

Travis, J. 1981. The effect of staining on the growth of *Hyla gratiosa* tadpoles. Copeia 1981:193-196.

Turner, F. B. 1960. Population structure and dynamics of the western spotted frog, *Rana p. pretiosa* Baird and Girard, in Yellowstone Park, Wyoming. Ecological Monographs 30:251-278.

Twitty, V. C. 1966. Of scientists and salamanders. Freeman, San Francisco.

Ussher, G. T. 1999. Method for attaching radio transmitters to medium sized reptiles: trials on tuatara (*Sphenodon punctatus*). Herpetological Review 30:151-153.

Vinegar, M. B. 1975. Life history phenomena in two populations of the lizard *Sceloporus undulatus* in southwestern New Mexico. American Midland Naturalist 93:388-402.

Waichman, A. V. 1992. An alphanumeric code for toe clipping amphibians and reptiles. Herpetological Review 23:19-21.

Ward, F. P., C. J. Hohmann, J. F. Ulrich, and S. E. Hill. 1976. Seasonal microhabitat selections of spotted turtles (*Clemmys guttata*) in Maryland elucidated by radioisotope tracking. Herpetologica 32:60-64.

Watson, J. W., K. R. McAllister, and D. J. Pierce. 2003. Home ranges, movements, and habitat selection of Oregon spotted frogs (*Rana pretiosa*). Journal of Herpetology 37:292-300.

Watters, T. S. and L. B. Kats. 2006. Longevity and breeding site fidelity in the California newt (*Taricha torosa*): a long-term study showing the efficacy of PIT tagging. Herpetological Review 37:151-152.

Weary, G. C. 1969. An improved method of marking snakes. Copeia 1969:854-855.

Weatherhead, P. J., and F. W. Anderka. 1984. An improved radio transmitter and implantation technique for snakes. Journal of Herpetology 18:264-269.

Weick, S. E., M. G. Knutson, B. C. Knights, and B. C. Pember. 2005. A comparison of internal and external radio transmitters with northern leopard frogs (*Rana pipiens*). Herpetological Review 36:415-421.

Wells, K. D., and R. A. Wells. 1976. Patterns of movement in a population of the slimy salamander, *Plethodon glutinosis*, with observations on aggregations. Herpetologica 32:156-162.

Wengert, G. M. and M. W. Gabrial. 2006. Using chin spot patterns to identify individual mountain yellow-legged frogs. [Abstract]. Northwestern Naturalist 87(2):192.

Whitaker, R. 1978. Permanent marking systems for crocodilians. Journal of the Bombay Natural History Society 75:496.

Whitford, W. G., and M. Massey. 1070. Responses of a population of *Ambystoma tigrinum* to thermal and oxygen gradients. Herpetologica 26:372-376.

Wilson, D. S. 1994. Tracking small animals with thread bobbins. Herpetological Review 25:13-14.

Windmiller, B. 1996. Tracking techniques useful for field studies of anuran orientation and movement. Herpetological Review 27:13-15.

Winne, C. T., J. D. Willson, K. M. Andrews and R. N. Reed. 2006. Efficacy of marking snakes with disposable medical cautery units. Herpetological Review 37(1):52-54.

Wisniewski, P. J., L. M. Paull, D. G. Merry, and F. M. Slater. 1980. Studies on the breeding migration and intra-migratory movements of the common toad (*Bufo bufo*) using Panjet dye- marking techniques. British Journal of Herpetology 6:71-74.

Woodbury, A. M. 1956. Uses of marking animals in ecological studies: marking amphibians and reptiles. Ecology 37:670-674.

Woodbury, A. M., and R. Hardy. 1948. Studies of the desert tortoise, *Gopherus agassizii*. Ecological Monographs 18:145-200.

Woolley, P. 1962. A method of marking salamanders. Missouri Speleology 4:69-70.

Woolley, H. P. 1973. Subcutaneous acrylic polymer injections as a marking technique for amphibians. Copeia 1973:340-341.

Zwickel, F. C., and A. Allison. 1983. A back marker for individual identification of small lizards. Herpetological Review 14:82.

Society Publications

PUBLICATIONS OF THE
SOCIETY FOR THE STUDY OF AMPHIBIANS AND REPTILES

SOCIETY PUBLICATIONS may be purchased from:
Breck Bartholomew, Publications Secretary
P.O. Box 58517 Salt Lake City, Utah 84158-0517, USA
Telephone and fax: area code (801) 562-2660
E-mail: ssar@herplit.com Web: http://www.herpsoc.org

Make checks payable to "SSAR" Overseas customers must make payment in USA funds using a draft drawn on American banks or by International Money Order. All persons may charge to MasterCard or VISA (please provide account number and expiration date).

Shipping and Handling Costs
• *Shipments inside the USA*: Shipping costs are in addition to the price of publications. Add an amount for shipping of the first item ($4.00 for a book costing $15.00 or more or$3.00 if the item costs less than $15.00) plus an amount for any additional items ($3.00 each for books costing over $15.00 and $2.00 for each item costing less than $15.00).
• *Shipments outside the USA*: Determine the cost for shipments inside USA (above) and then add 6% of the total cost of the order.
• *Large prints* (marked *): Shipments only inside the USA, add $5.00 for any quantity; regrettably, prints cannot be shipped outside the USA due to their large size.

CONTRIBUTIONS TO HERPETOLOGY
Book-length monographs, comprising taxonomic revisions, results of symposia, and other major works. Pre-publication discount to Society members. Missing volumes are out-of-print and no longer available from the SSAR.

Vol. 2. *The Turtles of Venezuela,* by Peter C. H. Pritchard and Pedro Trebbau. 1984. Covers half of the turtle species of South America. 414 p., 48 color plates (25 watercolors by Giorgio Voltolina and 165 photographs of turtles and habitats) measuring 8.5 x 11 inches, keys, 16 maps. Regular edition, clothbound $50.00; patron's edition, two leatherbound volumes in cloth-covered box, signed and numbered by authors and artist $300.00.

Vol. 6. *Snakes of the* Agkistrodon *Complex: A Monographic Review*, by Howard K. Gloyd and Roger Conant. 1990. Comprehensive treatment of four genera: *Agkistrodon, Calloselasma, Deinagkistrodon,* and *Hypnale*. Includes nine supplementary chapters by leading specialists. 620 p., 33 color plates (247 photographs of snakes and habitats), 20 uncolored plates, 60 text figures, keys, 6 charts, 28 maps. Clothbound $75.00. (*Also*: separate set of the 33 plates, in protective wrapper. $30.00; limited-edition print of the book's frontispiece illustrating snakes of all four genera, from watercolor by David M. Dennis, and signed by Dr. Conant and the artist $25.00.)

Vol. 7. *The Snakes of Iran*, by Mahmoud Latifi. 1991. Review of the 60 species of Iranian snakes, covering general biology, venoms, and snake bite. Appendix and supplemental bibliography by Alan E. Leviton and George R. Zug. 167 p., 22 color plates of snakes (66 figures), 2 color relief maps, 44 species range maps. Clothbound $25.00.

Vol. 9. *Herpetology: Current Research on the Biology of Amphibians and Reptiles.* 1992. Proceedings of the First World Congress of Herpetology (Canterbury, 1989), edited by Kraig Adler and with a foreword by H.R.H. Prince Philip, Duke of Edinburgh. Includes the plenary lectures, a summary of the congress, and a list of delegates with their current addresses. 225 p., 28 photographs. Clothbound $28.00.

Society Publications

Vol. 11. *Captive Management and Conservation of Amphibians and Reptiles*, by James B. Murphy, Kraig Adler, and Joseph T. Collins (eds.). 1994. Includes chapters by 70 leading specialists. Foreword by Gerald Durrell. 408 p., 35 photographs, 1 color plate. Clothbound $58.00.

Vol. 12. *Contributions to West Indian Herpetology*, by Robert Powell and Robert W. Henderson (eds.). 1996. Results of a Society-sponsored symposium, including research chapters by 59 authors and a checklist of species with complete citations. Foreword by Thomas W. Schoener. 457 p., 28 photographs, 70 color photographs, index. Clothbound $60.00.

Vol. 13. *Gecko Fauna of the USSR and Contiguous Regions*, by Nikolai N. Szczerbak and Michael L. Golubev. 1996. Covers the systematics, natural history, and conservation of the gecko fauna of the former Soviet Union and related species in surrounding regions from Mongolia through Pakistan, the Middle East, and northern Africa. 245 p., 24 colored and 60 black-and-white photographs, spot distribution maps, bibliography, index. Clothbound $48.00.

Vol. 14. *Biology of the Reptilia, vol. 19 (Morphology G)*, by Carl Gans and Abbot S. Gaunt (eds.). 1998. Chapters by 11 authors cover the major organs situated in the coelom: lungs, heart, liver, and spleen. 660 p., 145 figures, indices. Clothbound $58.00. (A complete list of the earlier volumes in this series [vols. 1-18], with names of publishers, is given in the book.)

Vol. 15. *The Lizards of Iran*, by Steven C. Anderson. 1999. Comprehensive summary, including systematics, natural history, and distribution. 450 p., 103 maps, 190 color photographs of lizards and habitats. Clothbound $65.00.

Vol. 16. *Slithy Toves: Illustrated Classic Herpetological Books at the University of Kansas in Pictures and Conversations*, by Sally Haines. 2000. A treasure trove of some of the finest illustrations of amphibians and reptiles ever produced, dating from the 16th to early 20th centuries. 190 p., 84 color photographs. Stiff paper cover $60.00.

Vol. 17. *The Herpetofauna of New Caledonia*, by Aaron M. Bauer and Ross A. Sadlier. French translations by Ivan Ineich. 2000. 322 p., 47 maps, 63 figures, 189 color photographs of animals and habitats. Clothbound $60.00.

Vol. 18. *The Hylid Frogs of Middle America*, expanded edition, by William E. Duellman. 2001. Review of the 165 hylid species from Mexico through Panama, with paintings by David M. Dennis. Foreword by David B. Wake. 1180 p., 443 figures and maps, 94 plates (46 in color). Clothbound in 2 volumes $125.00. (*Also*: separate set of the 46 color plates, in protective wrapper $45.00.)

Vol. 19. *The Amphibians of Honduras*, by James R. McCranie and Larry David Wilson. 2002. Comprehensive summary of 116 species, including systematics, natural history, and distribution. Foreword by Jay M. Savage. About 635 p., 126 figures, 33 tables, 154 color photographs of animals and habitats. Clothbound $60.00.

Vol. 20. *Islands and the Sea: Essays on the Herptological Exploration in the West Indies*, by Robert W. Henderson and Robert Powell (eds.). 2003. A collection of essays from 30 herpetologists on their experiences in the West Indies. 312 p,, 316 photos, 14 maps. Clothbound $48.00

FACSIMILE REPRINTS IN HERPETOLOGY

Exact reprints of classic and important books and papers. Most titles have extensive new introductions by leading authorities. Prepublication discount to Society members. Missing volumes are out-of-print and no longer available.

ANDERSON, J. 1896. *Contribution to the Herpetology of Arabia*. Introduction and new checklist of Arabian amphibians and reptiles by Alan E. Leviton and Michele L. Aldrich. 160 p., illus. (one plate in color), map. Clothbound $25.00.

BARBOUR, T. and C.T. RAMSDEN. 1919. *The Herpetology of Cuba*. Introduction by Rodolfo Ruidal. 200p., 15 plates. Clothbound. $55.00

BOURRET, R. 1941. *Les Tortues de l'Indochine*. Introduction by Indranel Das. 250 p. 48 uncolored and 6 colored plates. Clothbound. $65.00

Society Publications

COPE, E.D. 1864. *Papers on the Higher Classification of Frogs*. Reprinted from the Proceedings of the Academy of Natural Sciences of Philadelphia and Natural History Review. 32 p. $3.00.

—. 1871. *Catalogue of Batrachia and Reptilia Obtained by McNiel in Nicaragua; Catalogue of Reptilia and Batrachia Obtained by Maynard in Florida*. 8 p. $1.00.

COWLES, R. B., and C. M. BOGERT. 1944. *A Preliminary Study of the Thermal Requirements of Desert Reptiles*. With extensive review of recent studies by F. Harvey Pough. Reprinted from Bulletin of American Museum of Natural History. 52 p., 11 plates. Paper cover $5.00.

ESPADA, M. JIMENEZ DE LA. 1875. *Vertebrados del Viaje al Pacifico: Batracios*. A major taxonomic work on South American frogs. Introduction by Jay M. Savage. 208 p., 6plates, maps. Clothbound $20.00.

FAUVEL, A. A. 1879. *Alligators in China*. Original descriptionof *Alligator sinensis*, including classical and natural history. 42 p., 3 plates. Paper cover $5.00.

FERGUSON, W. 1877. *Reptile Fauna of Ceylon*. First comprehensive summary of the herpetofuana of Sri Lanka. Introduction by Kraig Adler. 48 p. $8.00

FITZINGER, L. 1826 & 1835. *Neue Classification der Reptilien* and *Systematische Anordnung der Schildkr ten*. Important nomenclatural landmarks for herpetology, including Amphibia as well as reptiles; world-wide in scope. Introduction by Robert Mertens. 110 p., folding chart. Clothbound $30.00.

FRANCIS, E.T.B. 1934. *Anatomy of the Salamander*. Forward by James Hanken and historical introduction by F.J. Cole. 465 p., 25 highly detailed plates, color plate. couthbound $60.00.

GRAY, J. E. 1825. *A Synopsis of the Genera of Reptiles and Amphibia*. Reprinted from Annals of Philosophy. 32 p. $3.00.

GRAY, J. E., and A. GUNTHER. 1845-1875. *Lizards of Australia and New Zealand*. The reptile section from "Voyage of H.M.S. Erebus and Terror," together with Gray's 1867 related book on Australian lizards. Introduction by Glenn M. Shea. 82 p., 20 plates (measuring 8.5 x 11 inches). Clothbound $20.00. (*Also*: set of the 20 plates, in protective wrapper $12.00.)

GUNTHER, A. 1885-1902. *Biologia Centrali-Americana. Reptilia and Batrachia*. The standard work on Middle American herpetology with 76 full-page plates measuring 8.5 x 11 inches (12 in color). Introductions by Hobart M. Smith, A. E. Gunther, and Kraig Adler. 575 p., photographs, maps. Clothbound $50.00. (*Also*: separate set of the 12 color plates, in protective wrapper $18.00.)

HOLBROOK, J. E. 1842. *North American Herpetology*. Five volumes bound in one. The classic work by the father of North American herpetology. Exact facsimile of the definitive second edition, including all 147 plates, measuring 8.5 x 11 inches (20 reproduced in full color). Introduction and checklists by Richard and Patricia Worthington and by Kraig Adler. 1032 p. Clothbound $60.00.

KIRTLAND, J. P. 1838. *Zoology of Ohio* (herpetological portion). 8 p. $1.00.

LECONTE, J. E. 1824-1828. *Three Papers on Amphibians*,from the Annals of the Lyceum of Natural History, NewYork. 16 p. $2.00.

MCLAIN, R. B. 1899. *Contributions to North American Herpetology* (three parts). 28 p., index. Paper cover $2.00.

ORBIGNY, A. D' [and G. BIBRON]. 1847. *Voyage dans l'Amerique Meridionale*. This extract comprises the complete section on reptiles and amphibians from this voyage to South America. 14 p., 9 plates measuring 8.5 x 11 inches. Paper cover $3.00.

PERACCA, M. G. 1882–1917. *The Life and Herpetological Contributions of Mario Giacinto Peracca (1861–1923)*. A collection of 64 herpetological titles, with descriptions of 22 amphibian species and 52 of reptiles. Introduction, annotated bibliography, and synopsis of taxa by Franco Andreone and Elena Gavetti. 550 p. Clothbound $55.00.

PETERS, W. 1838-1883. *The Herpetological Contributions of Wilhelm C. H. Peters (1815-1883)*. A collection of 174 titles, world-wide in scope, and including the herpetological volume in Peters' series, "Reise nach Mossambique." Biography, annotated bibliography, and synopsis of species by Aaron M. Bauer, Rainer Gunther, and Meghan Klipfel. 714 pages, 114 plates, 9 photographs, maps, index. Clothbound $75.00.

Society Publications

SCHMIDT, K. P., and G. K. NOBLE. 1919-1923. *Contributions to the Herpetology of the Belgian Congo.* Essential reference for the Congo rain forest and Sudanese savanna. Introductions by Donald G. Broadley and John C. Poynton. 780 pages, 141 photographs, maps, indices. Clothbound $65.00.

SHAW, G. 1802. *General Zoology, vol. 3: Amphibia.* Herpetological section from the first world summary of amphibians and reptiles in English. Introduction by Hobart M. Smith and Patrick David. 1014 p., 140 plates. Clothbound $75.00. .

SMITH, A. 1826-1838. *The Herpetological Contributions of Sir Andrew Smith.* A collection of 10 shorter papers including may descriptions of South African amphibians and repitles. Introduction by William R. Branch and Aaron M. Bauer. 83 p. Paper cover. $10.00.

SOWERBY,J.DeC., E.LEAR, and J.E. GRAY. 1872. *Tortoises, Terrapins, and Turtles Drawn From Life.* The finest atlas of turtle illustrations ever produced. Introduction be Ernest E. Williams. 26 p., 61 full-paged plates. Cothbound. $40.00

STEJNEGER, L. 1907. *Herpetology of Japan and Adjacent Territory.* Introduction by Masafumi Matsui. Also covers Taiwan, Korea, and adjacent China and Siberia. 684 pages, 35 plates, 409 text figures, keys, index. Clothbound $58.00.

TSCHUDI, J. J. VON. 1838. *Classification der Batrachier.* A major work in systematic herpetology, with introduction by Robert Mertens. 118 p., 6 plates. Paper cover $18.00.

—. 1845. *Reptilium Conspectus.* New reptiles and amphibians from Peru. 24 p. $2.00.

VANDENBURGH, J. 1914. *The Gigantic Land Tortoises of the Galapagos Archipelago.* The most extensive review of Galapagos tortoises. Foreword by Peter C. H. Pritchard. 290 pages, 205 photographs, maps, index. Clothbound $55.00.

WAITE, E. R. 1929. *The Reptiles and Amphibians of South Australia.* Introduction by Michael J. Tyler and Mark Hutchinson. 282 p., color plate, portrait, 192 text figures including numerous photographs. Clothbound $35.00.

WRIGHT, A. H., and A. A. WRIGHT. 1962. *Handbook of Snakes of the United States and Canada, Volume 3, Bibliography.* Cross-indexed bibliography to Volumes 1 and 2. 187 p. Clothbound $18.00.

JOURNAL OF HERPETOLOGY

The Society's official scientific journal, international in scope. Issued quarterly as part of Society membership. All numbers are paperbound as issued, measuring 7 x 10 inches.
Volumes 34-39 (2000-2005), four numbers in each volume, $9.00 per single number.
All previous volumes and numbers are out-of-print
Cumulative Index for Volumes 1-10 (1968-1976), 72 pages, $8.00.

HERPETOLOGICAL REVIEW AND H.I.S.S. PUBLICATIONS

The Society's official newsletter, international in coverage. In addition to news notes and feature articles, regular departments include regional societies, techniques, husbandry, life history, geographic distribution, and book reviews. Issued quarterly as part of Society membership or separately by subscription. All numbers are paperbound as issued and measure 8.5 x 11 inches. In 1973, publications of the Herpetological Information Search Systems (*News-Journal* and *Titles and Reviews*) were substituted for *Herpetological Review*; content and format are the same.
Volumes 31-36 (2000-2005), four numbers in each volume, $6.00.
All previous volumes and numbers are out-of-print
Cumulative Index for Volumes 1-7 (1967-1976), 60 pages, $5.00.
Cumulative Index for Volumes 1-17 (1967-1986), 90 pages, $8.00.
H.I.S.S. Publications: News-Journal, volume 1, numbers 1-6, and Titles and Reviews, volume 1, numbers 1-2 (all of 1973- 1974), complete set, $10.00.
Index to Geographic Distribution Records for Volumes 1-17 (1967-1986), including H.I.S.S. publications, 44 pages, $6.00.

Society Publications

CATALOGUE OF AMERICAN AMPHIBIANS AND REPTILES
Loose-leaf accounts of taxa (measuring 8.5 x 11 inches) prepared by specialists, including synonymy, definition, description, distribution map, and comprehensive list of literature for each taxon. Covers amphibians and reptiles of the entire Western Hemisphere. Issued by subscription. Individual accounts are not sold separately.

CATALOGUE ACCOUNTS:
   Complete set: Numbers 1-800, $450.00.
   Partial sets: Numbers 1-190, $65.00.
   Numbers 191-410, $75.00.
   Numbers 411-800, $320.00.
INDEX TO ACCOUNTS 1-400: Cross-referenced, 64 pages, $6.00;
   accounts 401-600: Cross-referenced, 32 pages, $6.00.
IMPRINTED POST BINDER: $35.00. (*Note*: one binder holds about 200 accounts.)
SYSTEMATIC TABS: Ten printed tabs for binder, such as "Class Amphibia," "Order Caudata," etc., $6.00 per set.

HERPETOLOGICAL CONSERVATION
A new series of book-length monographs, including symposia, devoted to all aspects of the conservation of amphibian and reptiles. Prepublication discount to Society members.

Vol. 1. *Amphibians in Decline: Canadian Studies of a Global Problem*, by David M. Green (ed.). 1997. Chapters by 52 authors dealing with population dispersal and fluctuations, genetic diversity, monitoring of natural populations, as well as effects of temperature, acidity, pesticides, UV light, forestry practices, and disease. 351 p., numerous photographs, figures, and tables, index. Paperbound $40.00

OTHER MATERIALS AVAILABLE FROM THE SOCIETY
The following color prints and brochures may be purchased from the Society. (*Extra postage required; see "shipping and Handling Costs" at the beginning of this list.)

*SILVER ANNIVERSARY COMMEMORATIVE PRINT.
   Full-color print (11.5 X 15.25 inches) of a Gila Monster (*Heloderma suspectum*) on natural background, from a watercolor by David M. Dennis. Issued as part of Society's 25th Anniversary in 1982. Edition limited to 1000. $6.00 each or $5.00 in quantities of 10 or more.
*WORLD CONGRESS COMMEMORATIVE PRINT. Full- color print (11.5 X 15.25 inches) of an Eastern Box Turtle (*Terrapene carolina*) in a natural setting, from a watercolor by David M. Dennis. Issued as part of SSAR's salute to the First World Congress of Herpetology, held at Canterbury, United Kingdom, in 1989. Edition limited to 1500. $6.00 each or $5.00 in quantities of 10 or more.
GRANTS AND AWARDS FOR HERPETOLOGISTS, by Joan C. Milam. 1997. A detailed listing, with descriptions and addresses, for about 100 research award programs. 106 p. $8.00.

Society Publications

Herpetological Circulars

Miscellaneous publications of general interest to the herpetological community. All numbers are paperbound, as issued. Prepublication discount to Society members. Missing numbers are out-of-print and no longer available from SSAR.

No. 8. A Brief History of Herpetology in North America Before 1900 by Kraig Adler. 1979. 40 p., 24 photographs, 1 map. $3.00.

No. 10. Vernacular Names of South American Turtles by Russell A. Mittermeier, Federico Medem, and Anders G. J. Rhodin. 1980. 44 p. $3.00.

No. 11. Recent Instances of Albinism in North American Amphibians and Reptiles by Stanley Dyrkacz. 1981. 36 p. $3.00.

No. 14. Checklist of the Turtles of the World with English Common Names by John Iverson. 1985. 14 p. $3.00.

No. 15. Cannibalism in Reptiles: A World-Wide Review by Joseph C. Mitchell. 1986. 37 p. $4.00.

No. 17. An Annotated List and Guide to the Amphibians and Reptiles of Monteverde, Costa Rica by Marc P. Hayes, J. Alan Pounds, and Walter W. Timmerman. 1989. 70 p., 32 figures. $5.00.

No. 18. Type Catalogues of Herpetological Collections: An Annotated List of Lists by Charles R. Crumly. 1990. 50 p. $5.00.

No. 21. Longevity of Reptiles and Amphibians in North American Collections (2nd ed.) by Andrew T. Snider and J. Kevin Bowler. 1992. 44 p. $5.00.

No. 22. Biology, Status, and Management of the Timber Rattlesnake (Crotalus horridus): A Guide for Conservation by William S. Brown. 1993. 84 p., 16 color photographs. $12.00.

No. 23. Scientific and Common Names for the Amphibians and Reptiles of Mexico in English and Spanish / Nombres Cient'ficos y Comunes en Ingles y Espa–ol de los Anfibios y los Reptiles de Mexico by Ernest A. Liner. Spanish translation by Jose L. Camarillo R. 1994. 118 p. $12.00.

No. 24. Citations for the Original Descriptions of North American Amphibians and Reptiles, by Ellin Beltz. 1995. 48 p. $7.00.

No. 26. Venomous Snakes: a Safety Guide for Reptile Keepers, by William Altimari. 1998. 28 p. $6.00.

No. 27. Lineages and Histories of Zoo Herpetologists in the United States, by Winston Card and James B. Murphy. 2000. 49 p., 53 photographs. $8.00.

No. 28. State and Provincial Amphibian and Reptile Publications for the United States and Canada, by John J. Moriarty and Aaron M. Bauer. 2000. 56 p. $9.00.

No. 29. Scientific and Standard English Names of Amphibians and Reptiles of North America North of Mexico, with Comments Regarding Confidence in Our Understanding, by the Committee on Standard English and Scientific Names (Brian I. Crother, chair). 2000 [2001]. 86 p. $11.00.

No. 30. Amphibian Monitoring in Latin America: a Protocol Manual/Monitoreo de Anfibios en America Latina:Manual de Protocolos, by Karen Lips, Jamie K. Reaser, Bruce E. Young, and Roberto Ibáñez. 2001 121 p. $13.00

No. 31. Herpetological Collecting and Collections Management. (Revised ed.) by John E. Simmons. 2002. 159 p. $16.00

No. 32. Conservation Guide to the Eastern Diamondback Rattlesnake Crotalus adamanteus. by Walter Timmerman and W. H. Martin. 2003. 64 pp. $13.00

No. 33. Chameleons: Johann von Fischer and Other Perspectives. by James B. Murphy. 2005. 123 pp. $13.00

No. 34 Synopsis of Helminths Endoparasitic in Sankes of the Untied States and Canada. by Carl H. Ernst and Evelyn M. Ernst. 2006. 90 pp. $9.00